## Geophysical Monograph Series

Including
**IUGG Volumes**
**Maurice Ewing Volumes**
**Mineral Physics Volumes**

# Geophysical Monograph Series

167 **Recurrent Magnetic Storms: Corotating Solar Wind Streams** *Bruce Tsurutani, Robert McPherron, Walter Gonzalez, Gang Lu, José H. A. Sobral, and Natchimuthukonar Gopalswamy (Eds.)*

168 **Earth's Deep Water Cycle** *Steven D. Jacobsen and Suzan van der Lee (Eds.)*

169 **Magnetospheric ULF Waves: Synthesis and New Directions** *Kazue Takahashi, Peter J. Chi, Richard E. Denton, and Robert L. Lysak (Eds.)*

170 **Earthquakes: Radiated Energy and the Physics of Faulting** *Rachel Abercrombie, Art McGarr, Hiroo Kanamori, and Giulio Di Toro (Eds.)*

171 **Subsurface Hydrology: Data Integration for Properties and Processes** *David W. Hyndman, Frederick D. Day-Lewis, and Kamini Singha (Eds.)*

172 **Volcanism and Subduction: The Kamchatka Region** *John Eichelberger, Evgenii Gordeev, Minoru Kasahara, Pavel Izbekov, and Johnathan Lees (Eds.)*

173 **Ocean Circulation: Mechanisms and Impacts—Past and Future Changes of Meridional Overturning** *Andreas Schmittner, John C. H. Chiang, and Sidney R. Hemming (Eds.)*

174 **Post-Perovskite: The Last Mantle Phase Transition** *Kei Hirose, John Brodholt, Thorne Lay, and David Yuen (Eds.)*

175 **A Continental Plate Boundary: Tectonics at South Island, New Zealand** *David Okaya, Tim Stern, and Fred Davey (Eds.)*

176 **Exploring Venus as a Terrestrial Planet** *Larry W. Esposito, Ellen R. Stofan, and Thomas E. Cravens (Eds.)*

177 **Ocean Modeling in an Eddying Regime** *Matthew Hecht and Hiroyasu Hasumi (Eds.)*

178 **Magma to Microbe: Modeling Hydrothermal Processes at Oceanic Spreading Centers** *Robert P. Lowell, Jeffrey S. Seewald, Anna Metaxas, and Michael R. Perfit (Eds.)*

179 **Active Tectonics and Seismic Potential of Alaska** *Jeffrey T. Freymueller, Peter J. Haeussler, Robert L. Wesson, and Göran Ekström (Eds.)*

180 **Arctic Sea Ice Decline: Observations, Projections, Mechanisms, and Implications** *Eric T. DeWeaver, Cecilia M. Bitz, and L.-Bruno Tremblay (Eds.)*

181 **Midlatitude Ionospheric Dynamics and Disturbances** *Paul M. Kintner, Jr., Anthea J. Coster, Tim Fuller-Rowell, Anthony J. Mannucci, Michael Mendillo, and Roderick Heelis (Eds.)*

182 **The Stromboli Volcano: An Integrated Study of the 2002–2003 Eruption** *Sonia Calvari, Salvatore Inguaggiato, Giuseppe Puglisi, Maurizio Ripepe, and Mauro Rosi (Eds.)*

183 **Carbon Sequestration and Its Role in the Global Carbon Cycle** *Brian J. McPherson and Eric T. Sundquist (Eds.)*

184 **Carbon Cycling in Northern Peatlands** *Andrew J. Baird, Lisa R. Belyea, Xavier Comas, A. S. Reeve, and Lee D. Slater (Eds.)*

185 **Indian Ocean Biogeochemical Processes and Ecological Variability** *Jerry D. Wiggert, Raleigh R. Hood, S. Wajih A. Naqvi, Kenneth H. Brink, and Sharon L. Smith (Eds.)*

186 **Amazonia and Global Change** *Michael Keller, Mercedes Bustamante, John Gash, and Pedro Silva Dias (Eds.)*

187 **Surface Ocean–Lower Atmosphere Processes** *Corinne Le Quèrè and Eric S. Saltzman (Eds.)*

188 **Diversity of Hydrothermal Systems on Slow Spreading Ocean Ridges** *Peter A. Rona, Colin W. Devey, Jérôme Dyment, and Bramley J. Murton (Eds.)*

189 **Climate Dynamics: Why Does Climate Vary?** *De-Zheng Sun and Frank Bryan (Eds.)*

190 **The Stratosphere: Dynamics, Transport, and Chemistry** *L. M. Polvani, A. H. Sobel, and D. W. Waugh (Eds.)*

191 **Rainfall: State of the Science** *Firat Y. Testik and Mekonnen Gebremichael (Eds.)*

192 **Antarctic Subglacial Aquatic Environments** *Martin J. Siegert, Mahlon C. Kennicut II, and Robert A. Bindschadler*

193 **Abrupt Climate Change: Mechanisms, Patterns, and Impacts** *Harunur Rashid, Leonid Polyak, and Ellen Mosley-Thompson (Eds.)*

194 **Stream Restoration in Dynamic Fluvial Systems: Scientific Approaches, Analyses, and Tools** *Andrew Simon, Sean J. Bennett, and Janine M. Castro (Eds.)*

195 **Monitoring and Modeling the *Deepwater Horizon* Oil Spill: A Record-Breaking Enterprise** *Yonggang Liu, Amy MacFadyen, Zhen-Gang Ji, and Robert H. Weisberg (Eds.)*

196 **Extreme Events and Natural Hazards: The Complexity Perspective** *A. Surjalal Sharma, Armin Bunde, Vijay P. Dimri, and Daniel N. Baker (Eds.)*

197 **Auroral Phenomenology and Magnetospheric Processes: Earth and Other Planets** *Andreas Keiling, Eric Donovan, Fran Bagenal, and Tomas Karlsson (Eds.)*

198 **Climates, Landscapes, and Civilizations** *Liviu Giosan, Dorian Q. Fuller, Kathleen Nicoll, Rowan K. Flad, and Peter D. Clift (Eds.)*

199 **Dynamics of the Earth's Radiation Belts and Inner Magnetosphere** *Danny Summers, Ian R. Mann, Daniel N. Baker, Michael Schulz (Eds.)*

200 **Lagrangian Modeling of the Atmosphere** *John Lin (Ed.)*

201 **Modeling the Ionosphere-Thermosphere** *Jospeh D. Huba, Robert W. Schunk, and George V Khazanov (Eds.)*

202 **The Mediterranean Sea: Temporal Variability and Spatial Patterns** *Gian Luca Eusebi Borzelli, Miroslav GaCiC, Piero Lionello and Paola Malanotte-Rizzoli (Eds.)*

Geophysical Monograph 203

# Future Earth—Advancing Civic Understanding of the Anthropocene

Diana Dalbotten
Gillian Roehrig
Patrick Hamilton
*Editors*

This Work is a co-publication between the American Geophysical Union and John Wiley & Sons, Inc.

WILEY

This Work is a co-publication between the American Geophysical Union and John Wiley & Sons, Inc.

## Published under the aegis of the AGU Books Board

Brooks Hanson, Director of Publications
Robert van der Hilst, Chair, Publications Committee
Richard Blakely, Vice Chair, Publications Committee

© 2014 by the American Geophysical Union, 2000 Florida Avenue, N.W., Washington, D.C. 20009
For details about the American Geophysical Union, see www.agu.org.
Published by John Wiley & Sons, Inc., Hoboken, New Jersey
Published simultaneously in Canada

No part of this publication may be reproduced, stored in a retrieval system, or transmitted in any form or by any means, electronic, mechanical, photocopying, recording, scanning, or otherwise, except as permitted under Section 107 or 108 of the 1976 United States Copyright Act, without either the prior written permission of the Publisher, or authorization through payment of the appropriate per-copy fee to the Copyright Clearance Center, Inc., 222 Rosewood Drive, Danvers, MA 01923, (978) 750-8400, fax (978) 750-4470, or on the web at www.copyright.com. Requests to the Publisher for permission should be addressed to the Permissions Department, John Wiley & Sons, Inc., 111 River Street, Hoboken, NJ 07030, (201) 748-6011, fax (201) 748-6008, or online at http://www.wiley.com/go/permission.

Limit of Liability/Disclaimer of Warranty: While the publisher and author have used their best efforts in preparing this book, they make no representations or warranties with respect to the accuracy or completeness of the contents of this book and specifically disclaim any implied warranties of merchantability or fitness for a particular purpose. No warranty may be created or extended by sales representatives or written sales materials. The advice and strategies contained herein may not be suitable for your situation. You should consult with a professional where appropriate. Neither the publisher nor author shall be liable for any loss of profit or any other commercial damages, including but not limited to special, incidental, consequential, or other damages.

For general information on our other products and services or for technical support, please contact our Customer Care Department within the United States at (800) 762-2974, outside the United States at (317) 572-3993 or fax (317) 572-4002.

Wiley also publishes its books in a variety of electronic formats. Some content that appears in print may not be available in electronic formats. For more information about Wiley products, visit our web site at www.wiley.com.

*Library of Congress Cataloging-in-Publication Data*

Future Earth : advancing civic understanding of the anthropocene / Diana Dalbotten, Patrick Hamilton, Gillian Roehrig, editors.
    pages   cm. – (Geophysical monograph series)
  Includes bibliographical references and index.
  ISBN 978-1-118-85430-3 (hardback)
1. Geological time.   2. Global environmental change.   3. Communication in science.   I. Dalbotten, Diana, 1959– editor of compilation.   II. Hamilton, Patrick, 1958– editor of compilation.   III. Roehrig, Gillian, 1968– editor of compilation.
  QE508.F88 2014
  304.2–dc23
                                                                                                                2013050492

Cover image: NASA Goddard Space Flight Center
Cover design by Modern Alchemy LLC

Printed in Singapore

10  9  8  7  6  5  4  3  2  1

# CONTENTS

**Contributors** .................................................................................................................... vii

**Preface** ............................................................................................................................. ix

**Acknowledgments** ........................................................................................................... xi

1. **Welcome to the Anthropocene**
   *Patrick Hamilton* ......................................................................................................... 1

2. **The Anthropocene and the Framework for K–12 Science Education**
   *Fred N. Finley* ............................................................................................................. 9

3. **Teacher Professional Development in the Anthropocene**
   *Devarati Bhattacharya, Gillian Roehrig, Anne Kern, and Melinda Howard* ................ 19

4. **Climate Literacy and Scientific Reasoning**
   *Shiyu Liu, Keisha Varma, and Gillian Roehrig* ........................................................... 31

5. **Evaluation and Assessment of Civic Understanding of Planet Earth**
   *Julie C. Libarkin* ........................................................................................................ 41

6. **Community-Driven Research in the Anthropocene**
   *Rajul E. Pandya* ......................................................................................................... 53

7. **Geoscience Alliance: Building Capacity to Use Science for Sovereignty in Native Communities**
   *Nievita Bueno Watts, Wendy Smythe, Emily Geraghty Ward, Diana Dalbotten, Vanessa Green, Mervyn Tano, and Antony Berthelote* ............................................................... 67

8. **New Voices: The Role of Undergraduate Geoscience Research in Supporting Alternative Perspectives on the Anthropocene**
   *Diana Dalbotten, Rebecca Haacker-Santos, and Suzanne Zurn-Birkhimer* ................. 77

9. **Shaping the Public Dialogue on Climate Change**
   *William Spitzer* .......................................................................................................... 89

10. **Opportunities for Communicating Ocean Acidification to Visitors at Informal Science Education Institutions**
    *Douglas Meyer and Bill Mott* ................................................................................... 99

11. **City-Wide Collaborations for Urban Climate Education**
    *Steven Snyder, Rita Mukherjee Hoffstadt, Lauren B. Allen, Kevin Crowley, Daniel A. Bader, and Radley M. Horton* ................................................................... 103

12. **On Bridging the Journalism/Science Divide**
    *Bud Ward* ................................................................................................................ 111

**Index** ............................................................................................................................ 121

Color plate section is located between pages 88 and 89.

# CONTRIBUTORS

**Lauren B. Allen**
Graduate Student
Learning Research & Development Center
University of Pittsburgh
Pittsburgh, Pennsylvania

**Daniel A. Bader**
Research Analyst
Center for Climate Systems Research
Columbia University
New York

**Antony Berthelote**
Hydrology Program Director
Salish Kootenai College Natural Resources Department
Pablo, Montana

**Devarati Bhattacharya**
Doctoral Candidate
STEM Education Center
Department of Curriculum and Instruction
University of Minnesota
St. Paul, Minnesota

**Nievita Bueno Watts**
Director of Academic Programs
Center for Coastal Margin Observation & Prediction
Institute of Environmental Health
Oregon Health & Sciences University
Portland, Oregon

**Kevin Crowley**
Professor
Learning Research & Development Center
University of Pittsburgh
Pittsburgh, Pennsylvania

**Fred N. Finley**
Associate Professor
STEM Education Center
Department of Curriculum and Instruction
University of Minnesota
St. Paul, Minnesota

**Emily Geraghty Ward**
Assistant Professor of Geology
Rocky Mountain College
Billings, Montana

**Vanessa Green**
Director of Higher Education and Diversity
Center for Coastal Margin Observation & Prediction
Institute of Environmental Health
Oregon Health & Sciences University
Portland, Oregon

**Rebecca Haacker-Santos**
SOARS Program Director, Head of Undergraduate Education
UCAR Science Education
University Corporation for Atmospheric Research
Boulder, Colorado

**Rita Mukherjee Hoffstadt**
Vice President, Education and Visitor Experience
San Antonio Children's Museum
San Antonio, Texas

**Radley M. Horton**
Associate Research Scientist
Center for Climate Systems Research
Columbia University
New York

**Melinda Howard**
Doctoral Student
Department of Curriculum and Instruction
University of Idaho–Coeur d'Alene

**Anne Kern**
Associate Professor, Science Education
Department of Curriculum and Instruction
University of Idaho–Coeur d'Alene

**Julie C. Libarkin**
Geocognition Research Lab
Department of Geological Sciences
Michigan State University
East Lansing, Michigan

**Shiyu Liu**
Department of Educational Psychology
University of Minnesota
Minneapolis, Minnesota

**Douglas Meyer**
Principal
Bernuth & Williamson
Washington, D.C.

**Bill Mott**
Director
The Ocean Project
Providence, Rhode Island

**Rajul E. Pandya**
Thriving Earth Exchange
American Geophysical Union
Washington, D.C.

University Corporation for Atmospheric Research
Boulder, Colorado

**Wendy Smythe**
K'ah Skaahluwaa
Center for Coastal Margin Observation & Prediction
Institute of Environmental Health
Oregon Health & Sciences University
Portland, Oregon

**Steven Snyder**
Executive Director
Reuben H. Fleet Science Center
San Diego, California

**William Spitzer**
Vice President
Programs, Exhibits, and Planning
New England Aquarium
Boston, Massachussetts

**Mervyn Tano**
President
International Institute for Indigenous Resource Management
Denver, Colorado

**Keisha Varma**
Assistant Professor
Department of Educational Psychology
University of Minnesota
Minneapolis

**Bud Ward**
Editor, The Yale Forum on Climate Change & The Media

**Suzanne Zurn-Birkhimer**
Associate Professor
Saint Joseph's College
Department of Mathematics
Rensselaer, Indiana

# PREFACE

When I was 10, I rushed into my family's living room Christmas morning to find a Sears and Roebuck telescope with my name on it. The perfect gift! At the time, I fantasized perpetually about exploring the unknown landscapes of Earth's nearest planetary neighbors. I promptly ventured out into the frigid December nights of 1968 and trained my new toy on Venus and Mars.

More than 40 years ago, scientists knew only a tiny fraction of what we know now about our nearest neighbors in the solar system, and so it was still possible for me to imagine a sultry world hidden beneath the dense clouds of Venus and envision ethereal creatures drifting through the thin atmosphere of Mars. As an adult, I now know better. Venus' supercharged hothouse atmosphere creates a surface environment of such intense heat and pressure as to make Death Valley at its most extreme seem like an alpine retreat. And the intensely cold and almost airless reality of Mars makes Antarctica seem like Tahiti by comparison.

So although each is beautiful and intriguing in its own way, Earth's closest neighbors in the solar system are remarkably inhospitable to human life. Also, even though they are close astronomically speaking, they are quite inaccessible in human terms. About 50 percent of all missions to Mars have failed.

Through the application of incredibly ingenious technologies and techniques, astronomers to date have confirmed 1074 planets orbiting other stars, with the count surely to rise rapidly in the near future. The scientific evidence increasingly points to planets being abundant in our galaxy, but these worlds beyond our solar system are ridiculously distant from Earth.

So here we are—working, playing, living, dying on a planet that is astonishingly conducive to life and now home to 7.1 billion people (with 9 billion expected by 2050). It is as if all of us were residing on a small ship in the midst of a vast ocean without end. No safe harbors are apparent, no verdant islands available for resupply and no other ships anywhere in sight if we begin taking on water. We are solely accountable for our circumstances.

We now live in a world being thoroughly reconfigured by human activity. Humans move more earth and rock annually that all rivers and glaciers combined. Humans fix more nitrogen than all microbial activity on the planet. Humans also currently appropriate nearly 40 percent of all terrestrial primary plant productivity. Humans are now such dominant agents of change that the term *Anthropocene* is used to describe this new geologic epoch in Earth's history.

Humans have initiated global changes that will reverberate for millennia, yet the misconception that the Earth is somehow too big and robust to be much influenced by human actions is a common one. We have the means to address the planetary challenges we collectively have set in motion. The future of Earth will be decided by human decision making, either by default or by design. What do we want our future Earth to be?

Patrick Hamilton, 2014

**Diana Dalbotten** *Director of Diversity and Broader Impacts, National Center for Earth-Surface Dynamics, St. Anthony Falls Laboratory, University of Minnesota, Minneapolis, Minnesota*

**Patrick Hamilton** *Program Director, Global Change Initiatives Science Museum of Minnesota, Saint Paul, Minnesota*

**Gillian Roehrig** *Associate Director and Associate Professor, STEM Education Center, University of Minnesota, Minneapolis, Minnesota*

# ACKNOWLEDGMENTS

Cover illustration: NASA Goddard Space Flight Center. This simulation used the Goddard Earth Observing System Model, Version 5 (GEOS-5) and the Goddard Chemistry Aerosol Radiation and Transport (GOCART) Model. GEOS-5 development is funded by NASA's Modeling, Analysis, and Prediction Program. An animation of the September 1, 2006 to March 17, 2007 aerosol simulation is available at http://sos.noaa.gov/Datasets/dataset.php?id=369.

This volume was made possible with the support of the National Science Foundation through the National Center for Earth-surface Dynamics (EAR-0120914) and the Future Earth Initiative (DRL-0741760). Any opinions, findings, and conclusions or recommendations expressed in this publication are those of the author(s) and do not necessarily reflect the views of the National Science Foundation.

# 1
## Welcome to the Anthropocene

### Patrick Hamilton*

*Whether we and our politicians know it or not, Nature is party to all our deals and decisions, and she has more votes, a longer memory, and a sterner sense of justice than we do.*—Wendell Berry

Over the past several decades, numerous independent lines of research conducted by thousands of researchers around the world have accumulated into a scientific body of evidence documenting that we all now live in a world being substantially reconfigured by human activity [*Pearce*, 2007]. As humans we have already set in motion global changes that will propagate for millennia [*Thompson et al.*, 2004]. Climate change as a result of releases of heat-trapping gases into the atmosphere is just one profound manifestation of this human-dominated planet that we all share. As of the writing of this chapter (January 2013), *Anthropocene* is not yet an official stratigraphic term, but it is being used with increasing frequency in both refereed publications and popular media to describe a new geologic epoch in Earth history [*Crutzen and Stoermer*, 2000].

The onset of the Anthropocene is a profound scientific realization, but it is not yet a widely shared new global paradigm. The dilemma of the Anthropocene is how evident it is to the geoscientific community while being little known at best to the general public. Public awareness of humanity as the dominant agent of global change significantly lags the scientific research, to the detriment of public policy efforts to stave off further undesirable planetary changes while preparing for those that are now unavoidable. It is not that most citizens are unwilling to accept the notion that humans now surpass natural processes in driving global change, but that many have not yet encountered it and those who have are often unsure of its significance. To geoscientists trained to consider global change in the context of vast stretches of geologic time, the planetary transformations that humans have set in motion are abrupt and without precedence in the geologic record [*Alley*, 2011]. No individual species in the history of this planet has come close to the domination of Earth now exercised collectively by all 7.1 billion humans, but to most people alive today our exceptional circumstances are merely the only reality they have ever known.

A multiplicity of innovations and solutions are needed at all scales of society for people to survive and thrive on a human-dominated planet. A more secure future is attainable if individuals and societies are open to new ways of achieving a high quality of life that reduces pressures on global environmental systems. A fundamental first step toward the creation of a more sustainable future is a broader societal realization that humanity has crossed a major threshold from being merely an inhabitant of Earth to being its leading architect and engineer. A number of articles and books have made strong and effective scientific cases for why humans no longer reside in the Holocene Epoch but rather in the Anthropocene one. This chapter, too, seeks to make the case for the Anthropocene but with a somewhat different tack. It will make the case for the Anthropocene in ways more in keeping with how those *not* trained in the geosciences experience the world around them.

---

*\*Program Director, Global Change Initiatives, Science Museum of Minnesota, Saint Paul, Minnesota*

---

*Future Earth—Advancing Civic Understanding of the Anthropocene, Geophysical Monograph 203*, First Edition.
Edited by Diana Dalbotten, Gillian Roehrig, and Patrick Hamilton.
© 2014 American Geophysical Union. Published 2014 by John Wiley & Sons, Inc.

In 2009, 29 international geoscientists collaborated on the creation of a consensus document intended to summarize the cumulative impact of humanity on the planet [*Rockström et al.*, 2009]. Nine key global systems were delineated of which seven were quantified. The purpose of this exercise was to formulate a scientific consensus on whether human activities were operating within or beyond the safe zones of our planet's biological, chemical, and physical systems. This chapter will not expound on the safe operating spaces for these nine planetary systems but rather will briefly survey major human influences on three prominent features of our planet: the 29 percent of Earth that is land, the other 71 percent that is covered by ocean, and the atmosphere that envelopes them both.

## 1.1. THE ANTHROPOCENE AND LAND

Buy land. They ain't making any more of the stuff.

—Will Rogers

Perhaps no transformation of our planet is more apparent to the casual human eye than land use. Cities are particularly striking human alterations of landscapes but in fact occupy a small portion of all land on the planet, only about 2 percent [*Henderson-Sellers and McGuffie*, 2012]. Agriculture has a giant global footprint. Forty percent of all ice-free land is committed already to growing our crops and raising our livestock. Croplands cover 1.53 billion hectares, which is an area comparable in size to all of South America, and pastures cover 3.38 billion hectares, which is an area similar in size to Africa [*Global Landscapes Initiative*, 2013].

Agriculture's dominance is evident on virtually any flight across every continent, except Antarctica. From where I live (Saint Paul, Minnesota, United States), window seats on flights reveal vast landscapes of cropland and pastureland for nearly 1,600 km (1,000 miles) in almost every direction, interrupted only by cities and towns, highways and railways, rivers and lakes, and occasional woodlands. The diverse mosaics of plant and animal communities that blanketed North America several hundred years ago have been replaced in enormous areas with a highly simplified plant ecology. In 2012 in the United States, just two plant species, corn and soybean, were planted on 70 million hectares, which an area larger than the state of California and comparable in size to entire countries (e.g. Sweden, Morocco, Papua New Guinea, or Uzbekistan).

So although North America encompasses biomes from tropical rainforests in the Yucatan of Mexico to Arctic tundra in northern Canada, much of the continent now lacks ecological integrity because many native plant and animal species face great difficulties in moving or migrating across human-engineered landscapes to reach other suitable habitats. For many plant and animal species, North America is no longer a landmass but ecologically more analogous to a vast archipelago of islands (some large, some tiny) separated from one another by seas of human-dominated landscapes of cities and transportation corridors, but especially croplands. This situation is true to greater or lesser degrees for all other continents on Earth, excluding Antarctica. Losses of terrestrial plant and animal species resulting from human activities are already high and are accelerating. Some extinction resulted from overharvesting, hunting species to extinction, or from introduced diseases or pests, but the vast majority of extinction has been and will continue to be as a result of loss of suitable habitat, unless we make major changes to agriculture.

The conversion of native lands to agriculture, particularly the species-rich tropical rainforests and savannas, must end. The gaps need to be closed between the yields on our best farms and low-yielding croplands with comparable climates and soils. Water, fertilizer, and energy must be consumed with much greater efficiency, especially given that 100 percent of freshwater resources are already dedicated to agriculture in many areas. Because much of the grain grown in the United States and Western Europe goes to fattening livestock, a reduction in eating meat could free up enormous quantities of calories for human consumption. Another huge gain in feeding the world could be achieved by reducing the 30 percent of food produced on the planet that currently is wasted in rich countries primarily by consumers and in poorer countries chiefly by failed crops, pest-infested stockpiles, or bad infrastructure and markets. Agriculture must be reimagined and reengineered on a massive scale if we are to stave off a mass extinction of plant and animal species and feed a world of nine billion people in 2050 [*Foley*, 2011].

## 1.2. THE ANTRHOPOCENE AND THE OCEAN

It is a curious situation that the sea, from which life first arose, should now be threatened by the activities of one form of that life.

—Rachel Carson

The ocean is home to some of the most remote and inaccessible ecosystems on the planet. Vast portions of the deep ocean remain unexplored and likely contain a high percentage of the organisms on Earth that have yet to be scientifically described. As terrestrial animals, the ocean that covers 71 percent of our planet can seem removed from our daily lives and thus largely safeguarded from significant human impact, although about *44 percent* of the world's population lives within about 160 km (100 miles) of the sea and most of the world's megacities (those with populations 10 million or more) with a cumulative total of more than 2.5 billion inhabitants are along coasts [*UN Atlas*, 2010].

These dense coastal concentrations of people heavily impact continental shelves and slopes. Coastal populations, furthermore, tend to be especially concentrated where rivers enter the ocean. These marine environments are impacted not only by coastal populations but also by the cumulative pollution load delivered by rivers and generated by human activities throughout watersheds that may drain millions of square kilometers of the interiors of continents. But human impacts now extend well beyond the coastal margins of continents.

Rapid advances in technology in recent decades have made it possible for humans to extract seafood and oil and gas from areas long considered inaccessible. Almost every fishery on the planet presently is being fished at or above what fishery scientists consider the maximum sustainable yield. Oil-drilling platforms now are capable of operating in 2 km (1.25 miles) of water to extract oil located below the seafloor. The rapid demise of summer sea ice in recent years likely means that the Arctic Ocean, once largely inhospitable to commercial activity, will see greatly increased pressure from the fishing, oil and gas extraction, and shipping industries in the near future. In addition to fishing and energy extraction industries, thousands of cargo ships now regularly ply intensively traveled shipping lines across the Atlantic, Pacific, and Indian Oceans with their concomitant noise, water, and air pollution.

The most pervasive human impact on the global ocean, however, is the result of human alterations of the atmosphere. Carbon dioxide ($CO_2$) and other heat-trapping gases released by human activities are enhancing the retention of heat near the Earth's surface. This process is raising temperatures over the continents, but the vast majority of the heat has been accumulating in the ocean. Scientists using different tools and different methods come up with differing ranges of upper ocean temperature increases since 1994, but all agree that the temperature of the upper ocean has been rising [*Lyman et al.*, 2010].

Ocean temperature increases have several implications. Water expands as it warms, so the absorption of heat by the ocean is contributing directly to sea level rise. Warming ocean water also appears to be melting the undersides of ice shelves along the coasts of the Antarctic Peninsula that help retard the flow of continental glaciers into the Southern Ocean. The increasing frequency of ice shelf collapses enables inland glaciers to flow more rapidly into the ocean, speeding up the delivery of ice and therefore water to the global ocean, further exacerbating sea level rise.

A warming ocean also means more heat energy potentially available for the formation of storms. The evidence is strong that the amount of water vapor in the atmosphere is increasing and causing precipitation to increase around the planet. The conversion of liquid water into water vapor and the eventual condensation of this vapor back into a liquid entail the absorption and release of enormous quantities of energy. This energy drives the global atmospheric circulation and spawns storm formation. The implications for the frequency and intensity of tropical cyclones in a warming world are still uncertain, but several billion people and trillions of dollars of infrastructure already are crowded onto coasts vulnerable to powerful storms.

To date, about 20 percent of the $CO_2$ released by human activities has been dissolving into the global ocean, which has had the beneficial effect of retarding the pace of global warming. The warming of the upper ocean, however, has significant long-term implications for the ability of the ocean to continue to absorb a substantial amount of anthropogenic $CO_2$. The amount of $CO_2$ that can remain in solution declines as water warms. This is evident in a glass of a carbonated beverage such as a soft drink. When pouring a glass, an effervescent froth quickly forms and then dissipates as $CO_2$ escapes from the liquid in response to the much lower outside ambient pressure as compared to inside the bottle. But after this initial release of $CO_2$, bubbles then continue to form on the bottom and sides of the glass as the liquid warms and its ability to store dissolved gases declines. As the upper ocean continues to warm, its ability to absorb $CO_2$ will decrease and with it, the ability of the ocean to ameliorate human-induced global warming. The potential exists, furthermore, that if ocean temperatures rise high enough, some of the $CO_2$ now in solution may instead be released back into the atmosphere, further exacerbating global climate change.

Human releases of $CO_2$ into the atmosphere are not only warming the global ocean but also changing its chemistry. When dissolved into water, $CO_2$ forms carbonic acid. Seltzer water, also known as club soda, is water with high concentrations of $CO_2$ dissolved into it. Pour a glass of seltzer and set it on a table. Over time much of the $CO_2$ will outgas and the pH of the water will rise as its acidity declines. As humans add $CO_2$ to the atmosphere, this process is operating in reverse on a global scale. $CO_2$ in the atmosphere is dissolving into the global ocean, producing carbonic acid, resulting in a decline in pH, and a rise in acidity of ocean waters everywhere.

Since the industrial revolution began, surface ocean pH is estimated to have dropped by slightly more than 0.1 units on the pH scale, from 8.179 to 8.069. Although this change might seem trivial, it represents a significant drop in pH because pH is measured on a logarithmic scale. This process of ocean acidification will continue and accelerate if human emissions of $CO_2$ continue their present upward trajectory. Scientists are scrambling to understand the implications of ocean acidification on marine life, but much already is known about how tiny changes in pH can have large impacts on the availability of the carbonate that many marine creatures, most notably corals, require to build their shells.

Human arterial blood must maintain a slightly alkaline pH of 7.41. With even slight variations from this level, there is virtually no part of the human body that will not suffer adverse and even life-threatening consequences. Humanity is in the process of altering the pH of the entire global ocean. It is impossible to know all of the consequences of this massive chemical experiment, but it can not end well for marine ecosystems everywhere if acidification is allowed to proceed unimpeded.

The only way to address ocean acidification is to dramatically reduce $CO_2$ emissions. Discussions regarding intentionally altering climate on a global scale to counteract the effects of human-caused increases in greenhouse gases (geoengineering) are increasing in scientific and political circles as apprehensions grow about the worsening effects of climate change. Although geoengineering the climate is theoretically possible, geoengineering the chemistry of the entire global ocean is not. In light of the vulnerability of the global ocean to excessive atmospheric $CO_2$ concentrations, geoengineering our way out of the escalating climate crisis may buy us some time, but on its own, it will ultimately fail to secure an enduring future for humanity.

## 1.3. THE ANTHROPOCENE AND THE ATMOSPHERE

> One thing we do know about the threat of climate change is that the cost of adjustment only grows the longer it's left unaddressed.
>
> —Jay Weatherill

$CO_2$ is an elusive gas. Colorless, odorless, and tasteless, it defies direct human experience. And it is present in the atmosphere in minute concentrations, only a few hundred parts per million. Oxygen and nitrogen, which together comprise 98 percent of the atmosphere, are by comparison about 500 and 2,000 times more abundant than $CO_2$.

Why does the increase of this trace gas from 280 parts per million (ppm) in the early 1800s to 392 ppm today cause such apprehension among atmospheric scientists? How can this gas have an outsized influence on the atmosphere? $CO_2$ and a few other trace gases in low concentrations in the atmosphere do what the vastly more abundant nitrogen and oxygen do not; they absorb and reradiate long-wave radiation (heat energy) emitted by the Earth's surface. The steady increase in atmospheric $CO_2$ means that more and more heat energy is not radiating out into space but rather is being retained, increasing the temperature of the lower atmosphere, the global ocean, and the continents.

Human activities, principally burning of fossil fuel and clearing of forests, are responsible for dramatic and ongoing increases in the concentrations of this trace gas in the atmosphere. The carbon in coal, oil, and natural gas has been locked away in geologic formations for millions, tens of millions, and in some instances, even several hundred million years. Humans now are extracting it from the earth, combining it with atmospheric oxygen in a chemical reaction that produces heat and light (energy) and then permitting the resulting molecules composed of one carbon atom and two oxygen atoms (O-C-O) to escape into the atmosphere. If present trends continue, the vast majority of fossil fuels formed over unbelievably long stretches of geologic time will be extracted and burned in only a few hundred years and humanity could push atmospheric $CO_2$ to levels not experienced on Earth in hundreds of thousands or even millions of years.

Vast quantities of $CO_2$ move between the atmosphere and biosphere on an annual basis. In particular, plants draw $CO_2$ out of atmosphere in the Northern Hemisphere when they photosynthesize during spring and summer, resulting in a small drop in atmospheric $CO_2$ during the growing season and then release this $CO_2$ back into the atmosphere as they die and decay during fall and winter. Atmospheric $CO_2$ then experiences a slight rebound. Atmospheric $CO_2$ had been remarkably stable at around 280 parts per million since the last Ice Age ended about 10,000 years ago despite enormous annual fluxes of $CO_2$ between the atmosphere and biosphere.

With the onset of the Industrial Revolution and large-scale burning of fossil fuels and clearing of land for agriculture, humans have been releasing $CO_2$ at a rate faster than physical, chemical, and biological processes can remove it from the atmosphere. A healthy human body is capable of metabolizing small, regular infusions of alcohol, but the physical and psychological manifestations of intoxication appear when alcohol is ingested faster than it can be detoxified by the liver. Humans are binging on carbon and so the concentration of $CO_2$ in the atmosphere is rising rapidly, altering Earth's energy balance, which in turn is causing the atmosphere to behave in increasingly aberrant ways.

The atmosphere is immensely complicated, but at its simplest conception, it is a heat engine. In thermodynamics, a heat engine is a system that converts heat energy into mechanical work by bringing a working substance from a high temperature state to a lower temperature state. An automobile's internal combustion engine is an example of a heat engine. Burning gasoline produces hot, rapidly expanding gases that drive the movement of the pistons and this motion gets translated to a drive shaft that then turns the wheels of the vehicle. The atmosphere too is a heat engine. Differences in solar energy input between the equator and the poles and between continents and oceans result in differences in temperature. Differences in temperature cause changes in pressure and the resulting winds transport air, water vapor, and heat energy around the planet.

The atmosphere is much more complex than an internal combustion engine but a basic principle is common to both systems: if additional energy is added, they must behave differently. With an automobile, stepping on the accelerator delivers more gasoline faster to the pistons, resulting in larger and more frequent explosions within the pistons and as a result, the car moves faster. Evidence is accumulating that the addition of more heat energy to the atmosphere has set in motion two fundamental changes in its behavior. The warming of the planet is accelerating the hydrological cycle, which is the process whereby water evaporates from the ocean, condenses into clouds, and then falls as precipitation only to resume the cycle all over again. The warming of the planet also is slowing the circulation of weather systems. Because of a variety of factors, polar regions are warming faster than equatorial ones, resulting in a lessening of temperature gradients and therefore pressure gradients and winds.

The retention of additional heat by the atmosphere is causing it to behave differently in ways apparent not only to climate scientists but also to casual observers. In Minnesota, winter temperatures are rising (especially nighttime lows), extreme rainfall events are increasing in frequency and intensity as are droughts, and the number of days per year with extremely high dew points is increasing. These changes have implications for all facets of the state's enterprises—agriculture, forestry, public and private infrastructure and investment—and are likely to continue and accelerate in coming decades because global emissions of $CO_2$ continue to rise. Agreements to radically reduce global $CO_2$ emissions in coming decades must be reached soon, lest changes to the global climate system be set in motion that will severely test the ingenuity of our highly intertwined and interdependent planetary human enterprise. But even if agreements are reached to mitigate $CO_2$ releases, we must acknowledge that sufficient quantities of this and other heat-trapping gases *already* have accumulated in the atmosphere to ensure climate change for decades and centuries into the future. Citizens individually and collectively need to take steps to adapt to climate change.

Although climate scientists have reached consensus on the basic fact that climate change is under way, the specific rates and degrees of change as well as the distribution of impacts are still incompletely understood. Interviews with 14 leading climate science experts suggest that over the next 20 years, research will be able to achieve only modest reductions in the degree of uncertainty about climate change and its impacts [*Zickfeld et al.*, 2010]. Uncertainty about climate change presents not only scientific challenges but also social, political, and economic quandaries [*Sarewitz et al.*, 2000]. Individuals and societies will need to make decisions about climate change with less-than-certain knowledge to guide them. Climate change will be unfair and unpredictable. Making decisions under conditions of increasing climate uncertainty will be the new normal.

## 1.4. THE ANTHROPOCENE AND HUMANITY

It was the best of times, it was the worst of times, it was the age of wisdom, it was the age of foolishness, it was the epoch of belief, it was the epoch of incredulity… .

—Charles Dickens, *A Tale of Two Cities*

The enormous pace of changes set in motion by human activities is without parallels in the geologic record. Never in Earth's 4.5-billion-year history has one species so thoroughly dominated the chemical, physical, and biological processes of the planet as humans do now, and we are in the process of not only transforming our planet but also ourselves. Earth now is home to the smartest, healthiest, wealthiest, best educated, most creative, innovative, and connected population in history.

• The IQs of people in many parts of the world have risen steadily since standardized IQ testing began in the 1930s. For example, a Dutch citizen of average intelligence in 1982 would have been considered a near genius in 1952 [*Flynn*, 2012]. The reasons for this increase in IQ continue to be researched and debated but regardless of its genesis, this phenomenon is a major global asset at a time when people everywhere individually and collectively must make more, better, and faster decisions about environmental issues.

• People, especially children, still needlessly die of easily preventable and treatable diseases, and AIDS has had devastating impacts on countries around the world, especially those in southern Africa. But lost in the sea of news about unnecessary death and suffering is the fact that global life expectancy at birth is now nearly 70 years of age, a dramatic increase from 53 years in 1960 [*World Bank*, 2013a]. This stunning improvement not only provides billions of individuals around the world with precious additional years of life but also ensures that many no longer die in the prime of their lives. All of the rest of us benefit from their accumulated experience, knowledge, and wisdom.

• The global economy generated nearly $70 trillion in goods and services in 2011. Although gross world product (GWP) has many deficiencies in terms of actually measuring the economic well-being and social welfare of people, it is nonetheless a startlingly indication of the sheer productivity of our collective global human enterprise. Income inequalities are rising in both developed and developing countries and debates are fierce in the United States and other countries about the appropriate role of governments in people's lives, but never in the history of humanity have so many people been so affluent. Twenty-seven percent of the global population was considered middle class in 2009. This figure is projected to rise to about 60 percent by the year 2030 [*Pezzini*, 2012].

• Education levels continue to rise steadily in most developed countries and are accelerating rapidly in many developing regions, especially east and south Asia. Around the world, 400 million people now hold college degrees—a population almost as large as that of the entire North American continent. Globally, the average number of years of schooling for people at least 15 years of age has more than doubled since 1950 to 7.76 years [*Barro and Lee*, 2010]. Economic output increases about 2 percent for each additional year in a population's schooling [*Wilson*, 2010]. Particularly striking is how the abilities of countries to address multiple problems, such as economic development, social welfare, environmental sustainability, and more, rise rapidly when girls gain opportunities to advance as far academically as possible [*Tembon and Fort*, 2008].

• Digital technology has set in motion an unprecedented transformation of the human experience. In 2012, 2.4 billion people (one out of three on the planet) had access to the Internet [*Internet World Stats*, 2012], and a year previously, mobile cellular telephone subscriptions worldwide exceeded six billion, which is nearly one for every person on Earth [*World Bank*, 2013b]. As of September 2012, Facebook had one billion members. Its rapid and continual growth ensures that Facebook's membership in the next year or two likely will eclipse the population of China, the world's most populous nation. The consequence of global Internet and phone connectivity has been an explosion in human collaboration, innovation, and creativity. Music, video, images, and ideas now regularly attract the attention of enormous global audiences almost instantly.

Humanity has the intellectual, educational, financial, creative, and collaborative assets necessary to manage the challenges of the Anthropocene. What has not yet arisen is an interconnected world view that we all are dependent on one another to keep Earth, which is the only planet known to be capable of supporting human life, a productive and secure shared home for ourselves and all the other species on which we depend. Global environmental consciousness is growing and maturing as our digital technology makes it increasingly possible for virtually anyone anywhere to become cognizant of the large-scale, adverse changes manifesting themselves both in their immediate environs and elsewhere in the world. But will an interconnected world view arise fast enough and gain sufficient political, economic, and social traction soon enough to set humanity on a new course?

## 1.5. THE RACE IS ON

Fasten your seatbelts, it's going to be a bumpy night.—Bette Davis as Margo Channing in *All About Eve*, 1950

We are in the midst of a tortoise and hare race. Humanity has set in motion global environmental changes that now will unfold over the coming years, decades, centuries, and millennia. Our challenge is to rapidly develop a global awareness of our circumstances and race ahead if we are to avert further undesirable environmental changes while adapting to those now unavoidable. The difference between our current global situation and the fable of the tortoise and the hare is that once we finally get into the race, we will be in it forever. Trying to manage a highly complex but finite planet and simultaneously balance the wants and needs of billions of people will be a never-ending responsibility for all current and future generations.

In coming years, it will become increasingly clear to a rapidly growing number of intelligent, well-educated, and connected people around the world that global environmental deterioration unfortunately will be a pervasive reality of their lives, given the large-scale changes that humans have already initiated and the current lack of a global consensus to address them. The social and psychological implications of this realization are unknown. Will citizens turn away from seemingly intractable global environmental change issues or recommit themselves to working to create a better world for future generations?

Many thoughtful, intelligent, and committed professionals in all walks of life are deeply engaged in the work of raising awareness that humanity now surpasses natural processes in driving global change. This book can give voice only to a few authors, but the ones in the following chapters were sought out because of the impressive range of their knowledge, expertise, and experience. Anthropocene education and outreach is by its nature multidisciplinary because it must encompass many fields of scientific inquiry and challenging to categorize because the most successful efforts are often eclectic in their approaches. A book, however, must have some form and structure, and so the editors of this publication have endeavored to group the contents according to the intended audiences and participants while acknowledging that this categorization is imperfect.

Chapters 2 and 3 investigate the teaching of Anthropocene concepts in K–12 formal educational settings, with emphases on standards and curriculum, teacher professional development, and assessment and evaluation. Chapters 4 and 5 explore how to enhance the scientific reasoning of the public to improve climate literacy and how we evaluate and assess the effectiveness of efforts to enhance civic understanding of the Anthropocene. Chapters 6 and 7 delve into efforts to engage with social groups often underserved by the geoscience community through participatory, community-based research efforts and through the building of research capacity on Native American reservations. Chapter 8 turns attention back to formal education but at the undergraduate level with deliberations on Anthropocene curriculum and faculty resources and the role of undergraduate research. Informal science education is the focus of Chapters 9 through 11 with an emphasis on the development and implementation of regional and national collaborations involving multiple institutions to reach large public audiences. And *Future Earth* concludes with Chapter 12, which reflects on bridging the divide between science and journalism.

This book seeks to be both informative and hopeful. The intention is that readers gain both knowledge and perspective and the encouragement to initiate, expand, and leverage other Anthropocene education efforts. People require hope if they are to persevere in any endeavor. They need to believe that their individual efforts when multiplied and leveraged by those of many others can achieve desired outcomes. Advancing civic understanding of the Anthropocene will require a careful and nuanced balancing act. People need to understand and appreciate the implications of the truly daunting global environmental challenges facing us all, but they also require evidence and examples that humanity can find a path to an enduring prosperity. We need to celebrate, promote, and publicize progress along this journey, while never forgetting that humanity has embarked on a new everlasting relationship with Earth, its only home.

## REFERENCES

Alley, R. (2011), *Earth: The Operators' Manual*, W. W. Norton & Company, NewYork.

Barro, R. J., and J. W. Lee (2010), A new data set of educational attainment in the world, 1950–2010, The National Bureau of Economic Research: NBER Working Paper No. 15902, April 2010, retrieved October 22, 2013, from http://www.nber.org/papers/w15902.

Crutzen, P. J., and E. F. Stoermer (2000), The "Anthropocene," *Global Change Newsletter*, 41: 12–13.

Flynn, J. R. (2012), *Are We Getting Smarter? Rising IQ in the Twenty-First Century*, Cambridge University Press, New York.

Foley, J. (2011), Can we feed the world and sustain the planet? *Sci. Am.*, November, 60–65.

Global Landscapes Initiative (2013), Institute on the environment, University of Minnesota, retrieved October 22, 2013, from http://environment.umn.edu/gli/.

Henderson-Sellers, A., and K. McGuffie (2012), *The Future of the World's Climate*, Elsevier, Waltham, Mass.

Internet World Stats (2012), Internet users in the world distribution by world regions 2012 Q2, retrieved October 22, 2013, from http://www.internetworldstats.com/stats.htm.

Lyman, J. M., S. A. Good, V. V. Gouretski, M. Ishii, G. C. Johnson, et al. (2010), Robust warming of the global upper ocean, *Nature*, 465, 334–337, retrieved October 22, 2013, from http://www.nature.com/nature/journal/v465/n7296/pdf/nature09043.pdf.

Pearce, F. (2007), *With Speed and Violence: Why Scientists Fear Tipping Points in Climate Change*, Beacon, Boston.

Pezzini, M. (2012), An emerging middle class, OECD Yearbook, retrieved October 22, 2013 from http://www.oecdobserver.org/news/fullstory.php/aid/3681/An_emerging_middle_class.html.

Rockström, J., W. Steffen, K. Noone, A. Persson, F. S. Chapin, et al. (2009), A safe operating space for humanity, *Nature*, 461, 472–475.

Sarewitz, D., R. A. Pielke, Jr., and R. Byerly, Jr. (Eds.) (2000), *Prediction: Science, Decision Making, and the Future of Nature*, Island Press, Washington, D. C.

Tembon, M. and L. Fort (Eds.) (2008), *Girls' Education in the 21st Century: Gender Equality, Empowerment and Economic Growth*, The World Bank, Washington, D.C.

Thompson, S. L., B. Govindasamy, A. Mirin, K. Caldeira, C. Delire, et al. (2004), Quantifying the effects of $CO_2$-fertilized vegetation on future global climate and carbon dynamics, *Geophys. Res. Letters*, 31 (23): L23211.

UN Atlas (2010), 44 percent of us live in coastal areas, retrieved October 22, 2013, from http://coastalchallenges.com/2010/01/31/un-atlas-60-of-us-live-in-the-coastal-areas/.

Wilson, D. (2010), College graduates to make global economy more productive: Chart of the day, Bloomberg.com, May 18, retrieved October 22, 2013, from http://www.bloomberg.com/news/2010-05-18/college-graduates-to-make-global-economy-more-productive-chart-of-the-day.html.

World Bank (2013a), Life expectancy at birth, retrieved October 22, 2013, from https://www.google.com/publicdata/explore?ds=d5bncppjof8f9_&met_y=sp_dyn_le00_in&tdim=true&dl=en&hl=en&q=global%20lifespan%20trends.

World Bank (2013b), Mobile cellular subscriptions, retrieved October 22, 2013, from http://search.worldbank.org/data?qterm=phone&language=EN&format=html.

Zickfeld, K., M. Granger Morgan, D. J. Frame, and D. W. Keith (2010), Expert judgments about transient climate response to alternative future trajectories of radiative forcing. *Proc. Natl. Acad. Sci. USA*, 107(28), 12451–12456.

# 2

# The Anthropocene and the Framework for K–12 Science Education

Fred N. Finley*

## 2.1. INTRODUCTION

*A Framework for K–12 Science Education: Practices, Crosscutting Concepts, and Core Ideas* [*National Research Council (NRC)*, 2012] is considered to be the most current and compelling guide available for science education across the various disciplines. The future of Earth Science education, and within that, teaching about the Anthropocene is likely to depend in large part on the ability of the Earth Science community to apply the framework. Applying the framework requires understanding the genesis, goals, and state of the current effort. Thus, an analysis and description of the essential components of the framework is provided first.

Analyses, discussion, and elaboration of recommendations about the use of the framework also will be needed. The United States has a history going back to the 1890s of developing well-grounded recommendations that never realized their potential in practice. The image that comes to mind is one of a life-giving rain storm over a parched landscape where one can see the rain falling but also sees it is not reaching the ground. The recommendations from many well-grounded framework-like reports have not reached the ground. A part of the problem is that we often do not consider what is implied by the recommendations and do not consider what elaborations are needed. This lack of elaboration results in overgeneralized and oversimplified applications.

Once we have analyzed, discussed, and elaborated on the basic recommendations, we perhaps can design,

research and evaluate, and revise standards, curriculum, instruction, and assessments in ways that will reach the ground. If we succumb to the temptations to overgeneralize and oversimplify recommendations related to the framework we will waste time, money, and public support for the teaching of the Earth Sciences. *The Framework for K–12 Science Education* deserves a better fate.

A chapter such as this can only begin a discussion of the elaborations that are needed and only for a part of what is included in the full document. The framework is too complex and extensive to try and accomplish more than that. This part of the discussion is primarily about standards and curriculum, that is, what the students are to know and what they are to be able to do. The discussion that is the last half of the chapter is organized under recommendations.

## 2.2. THE GENESIS AND GROUNDING OF THE FRAMEWORK

The *Framework for K–12 Science Education* was developed by the Board of Science Education within the Division of Behavioral and Social Sciences and Education of the NRC. The project members were drawn from the councils of the National Academy of Sciences, the National Academy of Engineering, and the Institute of Medicine. Other people responsible for the report were chosen for their special competencies to establish an appropriate balance across relevant sciences and areas of education.

The overall reasons for writing of the framework were the need for updating previous work; concerns about the effectiveness of science education; our position in the global economy; and the fact that a large number of states are considering adopting core standards in science. The most

*Associate Professor, STEM Education Center, Department of Curriculum and Instruction, University of Minnesota, St. Paul, Minnesota

*Future Earth—Advancing Civic Understanding of the Anthropocene, Geophysical Monograph 203*, First Edition.
Edited by Diana Dalbotten, Gillian Roehrig, and Patrick Hamilton.
© 2014 American Geophysical Union. Published 2014 by John Wiley & Sons, Inc.

immediate application of the framework is the development of the Next Generation Science Standards (NGSS) that were released during the week of April 8 to 12, 2013.

A review of the past 100 years or so of documents such as the framework indicate that it is a sophisticated, complex, and integrated set of ideas and proposals. The framework is supported by the most current research and best practices. The most recent and influential precursors to the framework are *Science for All Americans* [*Rutherford and Ahlgren*, 1990], *Benchmarks for Science Literacy* [*American Association for the Advancement of Science (AAAS)*, 1993], the *National Science Education Standards* [*NRC*, 1996], and the National Science Teachers Association (NSTA) Anchors Project [2009]; recent developments in the sciences and educational research; what has been learned from the various efforts to apply and implement the previous recommendations; and a substantive public review. In short, the recommendations are well grounded.

The framework is also a cautious document. The authors have taken care to acknowledge the great academic and practical complexity of designing and implementing truly effective science education in general, and Earth Science education in particular. They recognize that there are many interlocking national, state, and local components of our education system that have to be considered to develop sustainable effective science education programs. They also recognize that there is much that we do not know and need to know about students, teachers, and school systems, and thus, they call for a substantial research agenda. Finally, they recognize that the development and implementation of applications of the new standards is a massive undertaking that will require great resources and a sustained effort over a long time.

## 2.3. OVERALL GOAL

The goal statement for the framework is clear and succinct: The overarching goal of our framework for K–12 science education is to ensure that by the end of 12th grade, *all* students have some appreciation of the beauty and wonder of science; possess sufficient knowledge of science and engineering to engage in public discussions on related issues; are careful consumers of scientific and technological information related to their everyday lives; are able to continue to learn about science outside school; and have the skills to enter careers of their choice, including (but not limited to) careers in science, engineering, and technology [NRC, 2012, p. 1].

The goals of the framework must be kept at the forefront of our thinking as we attempt to include the Anthropocene in the K–12 curriculum.

## 2.4. CRITICAL FRAMEWORK ELEMENTS

The critical features of the framework are scientific and engineering Practices, Crosscutting Concepts, and Core Ideas. Each of these design specifications for standards and curricula are categories of the kinds of knowledge that need to be taught. Each category is then to be populated with the subject-specific Earth Science Practices, Crosscutting Concepts, and Core Ideas that also need to be taught. Although each category is presented as a distinctive design specification indicating what to teach, there is an overriding and essential requirement that "[t]o support students' meaningful learning in science and engineering, all three dimensions need to be integrated into standards, curriculum, instruction, and assessment" [NRC, 2012, p. 2]. In addition, the framework indicates that these dimensions must be integrated with each other and with carefully selected subject-specific knowledge.

Practices are critical aspects of how scientists and engineers do their work, what might be called the "inside" nature of science and engineering. The Practices are put forward as employed in all sciences; highly interdependent; evidence based; and centered on continuous and rigorous evaluation by critique, analysis, and argumentation within scientific communities. Each practice is presented with the full recognition that the practices are interdependent and used iteratively, are intimately related to the content of each science, and represent many different discipline-specific practices within each science. This is a fundamentally different view of the nature of science than is the idea that science is a simplistic single scientific method or form of inquiry. The framework presents the Practices as follows:

• Asking questions (for science) and defining problems (for engineering)
• Developing and using models
• Planning and carrying out investigations
• Analyzing and interpreting data
• Using mathematics and computational thinking
• Constructing explanations (for science) and designing solutions (for engineering)
• Engaging in argument from evidence
• Obtaining, evaluating, and communicating information

*Crosscutting Concepts* describe and represent the most basic ideas with which we can understand and interact with the natural and engineered worlds. These are among the most powerful intellectual tools we have. Crosscutting Concepts are pervasive in all sciences, represent the structures and functions of the content of specific sciences, and link each science to the others. "These concepts help provide students with an organizational framework for connecting knowledge from the various disciplines into a coherent and scientifically based view of the world" [NRC, 2012, p. 83].

The Crosscutting Concepts are:
- Patterns
- Cause and effect: Mechanism and explanation
- Scale, proportion, and quantity
- Systems and system models
- Energy and matter: Flows, cycles, and conservation
- Structure and function
- Stability and change.

*Core Ideas* are ideas from specific disciplines. They provide the organizational structure that relate specific ideas to each other and play a role in determining what new ideas are and are *not* deemed scientifically valid and credible. Core Ideas are essential to explanations of relevant sets of phenomena. Without core ideas, a discipline would be an undisciplined, disorganized collection of "bits and pieces" of knowledge and isolated practices. Core Ideas were used in the framework:

> [T]o avoid shallow coverage of a large number of topics and to allow more time for teachers and students to explore each idea in greater depth. Reduction of the sheer sum of details to be mastered is intended to give time for students to engage in scientific investigations and argumentation and to achieve depth of understanding of the core ideas presented. Delimiting what is to be learned about each core idea within each grade band also helps clarify what is most important to spend time on and avoid the proliferation of detail to be learned with no conceptual grounding [*NRC*, 2012, p. 11].

Core disciplinary ideas are specified for the Physical Sciences; Life Sciences; Earth and Space Sciences; and Engineering, Technology, and Applications of Science. Given the interdisciplinary nature of the Earth Sciences, all of these disciplines must be considered when considering the Anthropocene. However, only the Earth Science Core Ideas are considered here. The following two quotes indicate the overall approach:

> The essence of the core ideas is captured in the following two statements: Earth consists of a set of systems—atmosphere, hydrosphere, geosphere, and biosphere—that are intricately interconnected. These systems have differing sources of energy and matter cycles within and among them in multiple ways and on various time scales. Small changes in one part of one system can have large and sudden consequences in parts of other systems, or they can have no effect at all. Understanding the different processes that cause Earth to change over time (in a sense, how it "works") therefore requires knowledge of the multiple systems' interconnections and feedbacks [*NRC*, 2012, p. 170].

> [w]e begin at the largest spatial scales of the universe and move toward increasingly smaller scales and a more anthropocentric focus [NRC, 2012, p. 170].

The Core Ideas are:

ESS1: *Earth's Place in the Universe* (that) describes the universe as a whole and addresses its grand scale in both space and time. This idea includes the overall structure, composition, and history of the universe, the forces and processes by which the solar system operates, and Earth's planetary history [*NRC*, 2012, p. 170].

ESS2: *Earth Systems* (that) encompasses the processes that drive Earth's conditions and its continual evolution (i.e., change over time). It addresses the planet's large-scale structure and composition, describes its individual systems, and explains how they are interrelated. It also focuses on the mechanisms driving Earth's internal motions and on the vital role that water plays in all of the planet's systems and surface processes [*NRC*, 2012, p. 170].

ESS3: *Earth and Human Activity*, (that) addresses society's interactions with the planet. Connecting the ESS to the intimate scale of human life, this idea explains how Earth's processes affect people through natural resources and natural hazards, and it describes as well some of the ways in which humanity in turn affects Earth's processes [*NRC*, 2012, p. 170].

Under each of these categories of Core Ideas are three to five more specific categories of component concepts. For example, the categories under Earth's Systems are:

ESS2:   Earth's Systems
ESS2.A: Earth Materials and Systems
ESS2.B: Plate Tectonics and Large-Scale System Interactions
ESS2.C: The Roles of Water in Earth's Surface Processes
ESS2.D: Weather and Climate
ESS2.E: Biogeology [*NRC*, 2012, p. 171]

Under each category of the framework are indicators of the more specific ideas that are given by providing a basic question such as "*How do Earth's major systems interact?*" under ESS2B or "*How do the properties and movements of water shape Earth's surface and affect its systems?*" under ESS2.C [*NRC*, 2012, p. 184].

Each question is followed by what would be a good response. The responses range from one to seven paragraphs and include the even more specific ideas that are needed.

The framework also provides band end points:

> Band end points describe the developing understanding that students should have acquired by the ends of grades 2, 5, 8, and 12, respectively.... . These endpoints [also] indicate how this idea should be developed across the span of the K–12 years [*NRC*, 2012, p. 33].

For example, the Grade 8 band end under Earth Systems is:

> Plate tectonics is the unifying theory that explains the past and current movements of the rocks at Earth's surface and provides a framework for understanding its geological history. Plate movements are responsible for most continental and ocean floor features and for the distribution of most rocks and minerals within Earth's crust. Maps of ancient land and water patterns, based on investigations of rocks and fossils, make clear how Earth's plates have moved great distances, collided, and spread apart [*NRC*, 2012, p. 183].

## 2.5. RECOMMENDATIONS FOR DESIGNING STANDARDS AND CURRICULUM

One application of the framework is in designing standards and curriculum, that is, specifications of what students are to know and be able to do. Recommendations and accompanying discussions to that end are provided here as a starting point for further discussions. The first four recommendations are from the framework. These recommendations are followed by two that ask for elaborations of recommendations that are implicit or weak; two recommendations that were not part of the framework; a recommendation related to research; and one for more extended uses of the framework.

The first recommendation is to teach Crosscutting Concepts, Core Ideas, and Practices explicitly and have them coordinate with the teaching of the subject-specific concepts and practices. In the past, standards and curriculum development has tended to focus on teaching subject-specific concepts, has given some modest attention to specific scientific practices, and has left more general Crosscutting Concepts, Core Ideas, and Practices implicit at best.

Crosscutting Concepts represent key features of the sciences that cut across all sciences. Patterns are one such concept. All sciences engage in identifying patterns in phenomena. With respect to the Anthropocene, one example is that students can be taught explicitly the pattern of Hubbert Curves as they have been applied to human interactions with fossil fuel resources. The teaching of this pattern then can be extended to having students learn that this pattern is the same for all finite resources (e.g., phosphates and rare earths). Scale is another crosscutting concept that is used in many disciplines. Students can be taught about the relative long-term time scales of phenomena such as groundwater and landscape formation in contrast with the short time scales of human impacts. Teaching the concept of scale would be integrated with teaching the subject specific concepts for groundwater and landscape formation.

Core ideas are powerful principles that are needed to understand a particular science, in this case, the Anthropocene. An example would be teaching that matter and energy are transferred and transformed within and across the Earth's systems. Tracing the short- and long-term transfers and transformations of sunlight and human interactions with its various transfers and transformations would provide the context for teaching this Core Idea and subject- specific radiant energy interactions with the atmosphere, hydrosphere, and biosphere. Similarly, teaching students about short- and long-term cycling of matter within and across the Earth systems can be done using carbon as the matter that is being recycled. This would coordinate the explicit teaching of the core concept of cycling with the subject matter specific short- (especially human interactions with carbon) and long-term concepts related to carbon transfers and transformations.

The framework also requires the explicit teaching of practices such as engaging in argument from evidence. To do so effectively, we will need to decide exactly what forms of argument from evidence will be taught. For example, there are several different forms of argument used in studying the Anthropocene. Some practices are predictive, some retrodictive, some historical, some experimental, and some are based on modeling. Each will have to be taught in its own right using subject-specific cases such as predicting volcanic eruptions and associated human impacts; historical examinations of the impacts of population growth on water resources and soils; experimentation related to the effects of anthropogenic carbon releases; and modeling of the effects of rising sea levels.

The explicit teaching of Crosscutting Concepts, Core Ideas, and Practices is essential. We cannot expect students to somehow magically infer their existence, understand their intellectual power, and know how to use them with respect to new phenomena. There is little, if any, evidence that this kind of incredibly sophisticated learning—a move from novice to expert—occurs "automatically."

The second recommendation is that the Crosscutting Concepts, Practices, and Core Ideas be integrated with each other and with subject-specific concepts, and in the present case, concepts related to the Anthropocene. If the three design specifications rarely have been taught explicitly, they have almost never been taught in an integrated way. The extensive and complete integration that is used in the framework calls for an approach that is radically different from the traditional one. What is selected for teaching about the Anthropocene will have to be excellent and obvious exemplars of the Crosscutting Concepts, Practices, and Core Ideas and their interrelationships. For example, the teaching of the Core Ideas about the Earth and Human Activity such as climate change will require understanding Crosscutting Concepts such as scale (human and geologic time and spatial scales), cause and effect (effects caused by rapid anthropogenic releases of carbon into the atmosphere), and several kinds of scientific practices that have led to our current understanding of global warming (field-based observations, chemical and isotopic analyses of ice cores, experimentation, and modeling.)

The third overall recommendation from the framework is that the numbers of subject-specific concepts and practices be limited. This recommendation will have to be applied to teaching about the Anthropocene as well as all areas of the Earth Sciences. Responding to this recommendation also is severely challenging to tradition. Limiting how much we teach is made even more difficult because the framework intentionally expands the Earth Science curriculum even as it recognizes the necessity of reducing how much is taught because limited instructional time is available. Increasing the emphasis on the Anthropocene could increase the pressure on standards and curriculum to be even more selective. We cannot continue to ask our teachers to teach more and more, faster and faster, and expect our students to develop a functional knowledge of anything. Fortunately, the Anthropocene can and should have a special place in the Earth and Space Sciences because it is the most relevant to Earth and Human Activity.

One way to limit what is taught is to teach what serves *all* (or nearly all) of the stated framework goals directly, explicitly, and obviously. Virtually any ideas and practices from the Anthropocene could be shown to relate to one goal or another. Selecting something to teach that serves only one or two goals will severely overload the standards and curriculum, especially because the Anthropocene is only a part of the Earth Science standards and curriculum. What is to be taught must meet the set of goals by having students (1) appreciate the beauty and wonder of science; (2) engage in public discussions on related issues; (3) consume scientific and technological information carefully; (4) continue learning about science outside of and after school; and (5) develop the skills needed to enter careers of their choice[*NRC*, 2012, p. 1]. Many Earth Science phenomena meet these requirements—the beauty and critical importance of water and our lands and the challenges of what are called Earth hazards, such as earthquakes and volcanoes, all can be used to meet all of the goals. As with the first two recommendations, the teaching for all the goals will need to be done explicitly if it is to be effective.

Another way to select and limit the subject-specific Anthropocene concepts and practices is to consider the context in which the teaching will occur. This recommendation is more applicable to determining the curriculum than to determining standards. Standards are usually written for states or for countries. Curricula are designed for more local settings and thus there are greater possibilities for considering the particulars of a certain context.

Considering the local context limits what is to be taught because not all phenomena are present everywhere. In addition, local matters are typically more familiar and interesting to students. Considering the context has three parts. The first part is to select phenomena or problems that are local (e. g., water resources, soil conservation, mining, farming, and land use planning). We cannot teach it all so we may as well teach what is immediately and truly relevant. Begin with teaching about the natural systems: how the land or water or natural resources came into existence, the nature and materials of the systems, the processes that change the system, the ways in which the materials of the systems are transformed and transferred within and across systems, and the ways in which humans interact with the phenomena via their social systems. The second part is to teach about the social systems: local businesses, governing bodies, government agencies, nongovernmental organizations, and community members and to involve them in the teaching. Including those who are impacted or impacting the local phenomena can provide insights into how various social systems work, local community engagement opportunities, varying cultural perspectives, local knowledge, and information about careers. The third part is to have students apply what is learned in local contexts to other less familiar contexts in their region, nation, and other areas of the planet. Application of what is taught to other areas helps students understand the transferability of what they have learned and can help them understand similarities and differences in how other regions and cultures interact with similar phenomena. Many local phenomena are candidates such as frack sand mining, agricultural land-use practices, groundwater in Karst topographies, mining, beach erosion, and landslides,

The fourth recommendation is that the interdisciplinary nature of the Earth Sciences in general and Anthropocene in particular be fully recognized. The Earth Sciences are the epitome of interdisciplinarity. Nearly every area of the Earth Sciences references ideas from many other areas of the Earth Sciences usually in important ways. Thus, the Anthropocene cannot stand alone. Other aspects of the Earth Sciences, for example, the geologic and atmospheric history of the planet, are essential to teaching about it. Furthermore, understanding the Anthropocene requires understanding the traditional basics of the Earth Sciences such as the rock types and the rock cycle; water reservoirs and the hydrologic cycle; fossil fuel and mineral resource formation; sedimentation; glaciation; oceanography; plate tectonics; and Earth hazards.

An additional complication arises under the recommendation to consider the interdisciplinarity of the Anthropocene. Understanding the many subject-specific Anthropocene concepts and practices requires understanding concepts and practices from the Physical Sciences, Life Sciences, Mathematics, Engineering, and Social Sciences as well. The laws of conservation of matter and energy, gravitation, physical and chemical properties of elements and compounds, the biological necessities of life, evolution, and short-term biological carbon cycling are examples of concepts from other disciplines. In short, we have to consider the question, "What else does someone need to know?" to meet the requirements of the standards and curriculum. Thinking that the Earth Sciences alone can provide all of the underlying concepts from these fields is unrealistic. The teaching of Earth Sciences and the other related subjects will need to be coordinated.

The fifth recommendation is that the use of the Social Sciences be extended. The authors of the framework limited reference to the Social Sciences intentionally. Their reasons were related to the original charge to the committee, the large numbers of the relevant Social Sciences and their theories and methods, the commitment to using the previous recommendations in science education, and the great complexity associated with integrating Social Sciences. These are traditional reasons for not addressing the Social Sciences beyond the bare minimum. However, the limited use of the Social Sciences is perhaps the biggest challenge in using the framework to teach about the Anthropocene. By definition the Anthropocene requires that great attention be given to the interactions between natural and social systems. Given our huge and incredibly rapid influences on the planet, it seems impossible to teach about how the Earth systems have changed and are currently working within the Anthropocene without the Social Sciences. I cannot envision teaching about the current and future states of our atmosphere, hydrosphere, geosphere, and biosphere (more specifically, weather and climate change, our oceans and freshwater, soil and mineral resources, and biodiversity) without substantial reference to our local, national, and international social systems. In fact, I cannot understand how one can teach anything about the Anthropocene without consideration of our social systems.

What is almost certainly needed is an expanded and elaborated view of the Earth systems. The current framework references teaching about the Earth's systems, the geosphere, atmosphere, hydrosphere and biosphere. This is missing an absolutely critical proposition and a corollary. The proposition is that humans are part of and not separate from the biosphere. The corollary is that humans have created complex powerful social systems. If humans and our social systems are part of the biosphere, we must include ourselves as part of the Earth's systems. We have stepped off this planet and looked back at the planet that we occupy. When we ask how we can understand the planet and how it works, it seems that this can best be done in terms of interacting natural systems and the social systems. The social systems that need to be included are economics, politics, agriculture, transportation, business and industry, and government at local, national, and international levels, and culture at a minimum.

All of the framework committee's arguments for not including the Social Sciences are real concerns; however, the conclusion to limit the attention given to the Social Sciences is not valid, especially with respect to teaching about the Anthropocene. Instead of avoiding the aforementioned problems, they must be addressed. The work will be incredibly difficult and complex. This work cannot and should not be done just by Earth scientists and educators. What will be required is a truly cooperative interdisciplinary effort that would include social scientists and educators and their high-status institutions. Such an effort would give real credence to the framework's emphasis on interdisciplinarity. In the meantime, Earth scientists and science educators need to use the idea of the Earth as a set of interacting natural and social systems as much as possible and work out how to do so effectively under the current circumstances.

The sixth recommendation is that the use of the history, philosophy, and sociology of science be extended. The limited attention given to these fields is somewhat surprising. The authors indicate that:

> [s]cience is fundamentally a social enterprise, and scientific knowledge advances through collaboration and in the context of a social system with well-developed norms. Individual scientists may do much of their work independently or they may collaborate closely with colleagues. Thus, new ideas can be the product of one mind or many working together. However, the theories, models, instruments, and methods for collecting and displaying data, as well as the norms for building arguments from evidence, are developed collectively in a vast network of scientists working together over extended periods. As they carry out their research, scientists talk frequently with their colleagues, both formally and informally. They exchange emails, engage in discussions at conferences, share research techniques and analytical procedures, and present and respond to ideas via publication in journals and books. In short, scientists constitute a community whose members work together to build a body of evidence and devise and test theories [*NRC*, 2012, p. 27].

Although this quotation plus a mention of the roles of other scientists under the practice of engaging in argument from evidence would provide justification for teaching of the history and sociology of science, the ideas contained in the framework are not sufficiently prominent in the document to demand that they be used. The exclusion of many social practices and motivations of science such as competition for grants, first publication rights, financial gain, recognition among peers, collaboration, communications, sometimes vehement personal arguments, and even fraud overly sterilize the science the students would see by removing the deeply human aspects of the work. This presentation of science with so little about the human aspects of it included misrepresents the nature of science and may also decrease interest in the sciences.

A part of this recommendation to increase the use of the history, philosophy, and sociology of science is that the interactions between science and society be included as well. This is implied by the following quote—even though it receives almost no additional attention—

> [t]his (scientific) community and its culture exist in the larger social and economic context of their place and time and are influenced by events, needs, and norms from outside science, as well as by the interests and desires of scientists [*NRC*, 2012, p. 27].

Nearly all the references to practices are references to what happens "inside science." Only weak reference is made to what happens "outside science," that is, the interactions among scientific practices and society. Our social systems—economic, political, communications, value, moral, religious, and ethical—strongly influence what science is undertaken, how the science is done, and how influential it is in our lives. One might expect that the limited attention given to the human nature of scientists and the interactions with society is because the teaching of Earth Sciences has typically (as has the teaching of most sciences) excluded humans except for passing references to famous scientists. This has been done under the formerly accepted idea that science is absolutely objective, governed only by formal logic, and independent of our human limitations. Since the traditional objective, formal logic view of science is still at work, then the tradition has to change because there is little in studies of the nature of science over the past four or more decades that would support continuing the myth. Our students need to learn about science as a human endeavor and as the product of our wonder and curiosity, creativity, intellect, aspirations, social systems, and cultures. An important corollary to this recommendation is to expand the practices to include what people do with their scientific knowledge outside of the sciences. The practices that are included in the framework at this time are found by looking inside science. The strong interactions between the sciences and society along with the inclusion of social systems in the Earth Sciences seem to require that the recommended practices include teaching individuals to make short- and long-term personal and social decisions. We need to teach when and how science can and should be used in making such decisions. We also need to teach how to evaluate the validity of purportedly scientific communications, respond to those communications, and include science in our daily interactions with others. Finally, our students need to be taught and even experience Earth Science–related careers. One of the most compelling ways to engage students in the Earth Sciences is to have community members who work in Earth Science–related fields collaborate with the schools. The collaborations can range from guest presentations, to field trips to actual community-based projects with local businesses and industries, nongovernmental organizations, and government agencies. In short, the practices need to include our uses of knowledge outside of the sciences.

One promising way of teaching about science and society and the nature of science is the development and use of case studies. Earth Science is replete with cases about the human aspects of science. For example, past cases such as the Cope and Marsh Bone Wars and Wegner's development of the theory of continental drift plus current cases such as Climategate, the arguments about fracking, and groundwater issues such as occur in the Ogallala aquifer are all cases that provide context for not only the teaching about science and society and the nature of science but also about Crosscutting Concepts, Core Ideas,, Practices and subject-specific concepts as well. Cases are known to be engaging for students because the human dimensions are so intriguing.

The seventh recommendation is that standards and curriculum for the Anthropocene begin with descriptions of the Earth systems phenomena and then move on to learning to make warranted explanations, justified predictions, and defensible decisions. Starting with descriptions is critical. The framework calls for students to learn to make explanations. However, it does not seem to recognize the fact that students do not know even the basics of what materials and structures are on the planet. For example, they do not know the basic materials, structures, locations of the ocean floor or where the major faults, mountain ranges, volcanic regions, and tectonic plates are located. With respect to the human interactions with the planet, they do not know the history and distribution of human populations on the planet or basic patterns of resource extraction, consumption, and waste management. Rich descriptions of the planet are needed before explanations of "how things work." Otherwise we are essentially asking students to learn explanations about phenomena that in some real sense they have never seen.

Once rich descriptions are available, then the explanations of how things work, such as interacting systems in terms of Earth processes, transfers, and transformations of matter and energy and human interactions on different time scales, can be provided. Once explanations and predictions can be made, then students can be asked to identify what phenomena are problematic for humans, what options we have for ameliorating the problems, and conclude the curriculum with arguing for and making personal and social decisions that can improve the chances of a sustainable future for humans.

This last step provides a venue for considering controversies. Moving from description to explanation to prediction to recognizing problems to considering options to decision making avoids the often discouraging curricula that start with the problems that seem intractable and argues for "saving Mother Earth." The approach acknowledges that the Earth does not care what we do; it has been here long before we arrived, will be here long after we are gone, and is not sentient. The proposed approach argues that we do or at least should care and provides grounding in the Earth Sciences for considering alternative solutions to our problems.

The eighth recommendation is that three additional Core Ideas are probably needed. Two of these ideas, chaos and complexity, are critical in developing a current understanding of the Earth systems. The core premise of chaos theory is that small differences in initial conditions can have unpredictable, nonrepeating, nonlinear large impacts on dynamic systems over time [*Kellert*, 1993]. The core principle of complexity is that a complex system is one in which numerous independent elements continuously interact and spontaneously organize and reorganize themselves into more and more elaborate structures over time

[*Williams*, 1997]. The third idea is sustainability. Sustainability refers to using current natural resources in ways that meet the needs of the current generation but allow for meeting the needs of future generations without harming the future generations. Given that humans have innumerable chaotic impacts of varying magnitudes on the Earth systems, and a moral obligation to future generations, including these ideas as core concepts in standards and curriculum for the Anthropocene seems essential.

The final two recommendations are somewhat different in that they are about the framework overall and the importance of research in Earth Science education.

The first of these recommendations is taken from the chapter "Looking Toward the Future: Research and Development." Scientists, educators, and teachers must participate in extensive research and evaluation. Analytic studies that generate models of ways in which the Crosscutting Concepts, Core Ideas, Practices, subject-specific concepts and practices, the natures of the sciences, interactions between science and society, and the interactions of natural and social systems can be integrated will be essential to designing standards and curricula.

Little is known about teaching or about students learning Crosscutting Concepts, generalized Practices, and even intellectually powerful Core Ideas. Although we have a few studies of teaching and learning of a few subject-specific concepts in Earth Science, we have virtually no studies of what the framework calls "progressions of teaching and learning." We will need to ask questions about these matters, students' prior knowledge both separately and when integrated; changes in their knowledge when different teaching strategies and methods are used; and new assessment methods. Fortunately, a variety of educational theories and empirical studies related to learning and cognition, conceptual change, problem solving, scientific reasoning, the roles of social interactions in learning, cultural influences, and motivation are available for use in this research. Likewise a variety of qualitative and quantitative educational research methods that can be used separately or integrated are also available. We will need in-depth small-scale studies and larger-scale studies within and across different educational contexts, socioeconomic backgrounds, genders, and cultural groups. We cannot afford years of trial and error and cannot afford designing standards, curriculum, instruction, and assessment without learning from multiple investigations.

The second of these recommendations is that all who are interested in teaching about the Anthropocene read the framework in its entirety. As was indicated previously, this is a comprehensive and sophisticated document. Much more is said about "Realizing the Vision," especially with respect to Implementation: Curriculum, Instruction, Teacher Development, and Assessment; Equity and Diversity in Science and Engineering Education, Guidance for Standards Developers; and Looking Toward the Future: Research and Development.

## 2.6. CONCLUSION

Among the ten recommendations given in this chapter was the recommendation to teach far fewer topics better, thus confronting the criticism that science curricula in the United States tends to be "a mile wide and an inch deep." Unfortunately, what was proposed makes the challenge even more daunting by pointing out that the framework expands what is to be taught with respect to the Earth Sciences. The recommendation to emphasize the Anthropocene expands this even further. This chapter also indicates that we need to teach the Crosscutting Concepts, Practices, and Core Ideas explicitly in a way that integrates them with each other and with subject-specific concepts and practices, to include more about the nature of Earth Science, the interactions between science and society, and the interactions of natural and social systems and to be sure we include "what else" has to be learned from related Earth Science fields and from the physical sciences as well. If we take these recommendations seriously, we cannot teach all that would be good to know about the Anthropocene. In fact, given that we have little instructional time, we can teach few topics if we want students' knowledge to become functional in the short term and preferably in the long term.

We have to limit severely what we teach from Earth Science in general and the Anthropocene in particular and then squeeze as much instructional time out of the overall curriculum as we can at all levels—elementary, middle school, and high school—and in informal settings as well. When we do so, we have to recognize that every other school subject has claims to be made on instructional time and resources. We must have help from the Social Sciences in particular. Given the inclusion of the Social Sciences as part of the Earth systems, we probably can ask that they can contribute. Perhaps we also can ask for cooperation with those from the Physical Sciences and Life Sciences by having them teach underlying concepts and practices from their fields using phenomena that are drawn from the Earth Sciences. However, if we ask for collaboration, we must be sure that what we ask is congruent with their standards and curricula and recognize that they too do not have enough space in their curricula. When we request help we are really asking them to not teach something they usually teach and think is important. In return, we can we pick up on their standards and make a point of teaching "their" ideas and practices during our Earth Science teaching and how their ideas are used in our fields. These collaborations are likely to increase the relevancy of their curricula and ours simultaneously.

If we ask for collaborations, perhaps we can see not only day-to-day ways to cooperate, but also the development of a small number (one to three) of sustained three- to four-weeklong interdisciplinary modules or projects with an Anthropocene focus each year. Perhaps these units can be taught by more than one subject simultaneously, mostly within each standing classroom. The units would have to contribute to meeting the standards for each classroom as well as interdisciplinary goals. If this were done even once each year from across the K–12 years, students would have learned much more about interactions between natural and social systems than they do now.

This collaborative curriculum would be different but is proposed because it is relatively practical. The educational systems would have to be somewhat different, especially the ways we plan curricula. Interdisciplinary teams would be needed and teaching schedules would need to be modified to at least allow for joint planning if not co-teaching. New local curriculum, teaching, and assessment materials would be needed as well. However, wholesale revisions might not be required.

Large-scale revisions of the overall school curriculum to accommodate more Earth Science teaching are not at all practical. Large-scale revisions are unrealistic in part because no one is looking at the curriculum across the various school subjects. The basic structure of the curriculum has been in place for more than 100 years and education is still dominated by discipline-by-discipline teaching. Each subject is on its own and competing for space with every other subject. Large-scale revisions would require national and state standards that were coordinated across subjects; reallocations of the time spent on each school subject; the replacement of the great mass of current curriculum materials; local, state, and national assessments to be changed; extensive teacher education; strong governmental and administrative leadership; and innumerable new and different administrative practices.

Perhaps large-scale change is a goal to be pursued over years of sustained commitment. In the meantime, making small inroads, finding "easy intersections" that meet the standards of more than one subject, and finding a few times each year when the Earth Sciences and other subjects can be coordinated is feasible.

## REFERENCES

American Association for the Advancement of Science: Project 2061 (1993), Benchmarks for Science Literacy, retrieved October 22, 2013,from http://www.project2061.org/publications/bsl/online/index.php?txtRef=http%3A%2F%2Fwww%2Eproject2061%2Eorg%2Fpublications%2Fbsl%2Fdefault%2Ehtm%3FtxtRef%3D%26txtURIOld%3D%252Ftools%252Fbsl%252Fdefault%2Ehtm&txtURIOld=%2Fpublications%2Fbsl%2Fonline%2Fbolintro%2Ehtm.

Kellert, S. (1993), *In the Wake of Chaos: Unpredictable Order in Dynamical Systems Chicago*, University of Chicago Press, Chicago.

National Research Council (1996), *National Science Education Standards*, National Academy Press, Washington, D.C.

National Research Council (2012), *A Framework for K–12 Science Education: Practices, Crosscutting Concepts, and Core Ideas*, National Academy Press, Washington, D.C.

National Science Teachers Association. Project: Science Anchors (2009), *Moving Standards and Assessments Forward*. Accessed November 3, 2013, from http://science.nsta.org/enewsletter/anchors.pdf.

Rutherford, F. J., and A. Ahlgren (1990), *Science for All Americans*, Oxford University Press, New York.

Williams, G. P. (1997), *Chaos Theory Tamed*, Joseph Henry Press, Washington, D.C.

# 3

# Teacher Professional Development in the Anthropocene

### Devarati Bhattacharya[1], Gillian Roehrig[2], Anne Kern[3], and Melinda Howard[4]

The recently published *Framework for K–12 Science Education* [*National Research Council (NRC)*, 2012] presents a vision for the future of K–12 science education. This vision is drawn in part by arguments based on the premise that continued prosperity and progress within the United States depends on the development of more highly qualified STEM professionals [*National Academy of Sciences, National Academy of Engineering, and Institute of Medicine of the National Acadamies*, 2007; *President's Council of Advisors on Science and Technology*, 2010; *Congress Joint Economic Committee*, 2012]. However, the *Frameworks* are clear that "all students, not just those who pursue careers in science, engineering, or technology or those who continue on to higher education" [*NRC*, 2012, p. 9] need to be scientifically literate and develop the knowledge and skills to be able to,

> engage in public discussions on science-related issues, to be critical consumers of scientific information related to their everyday lives, and to continue to learn about science throughout their lives [*NRC*, 2012 p. 9]

The *Frameworks* strive to move this vision forward through the use of developmental learning progressions, focusing on a limited number of core ideas in science and engineering, and emphasizing that learning about science and engineering involves the integration of content and practices. Structurally, the *Frameworks* organize K–12 science education around three major dimensions:

• Scientific and engineering practices;
• Crosscutting concepts that unify the study of science and engineering through their common application across fields;
• Core ideas in four disciplinary areas: physical sciences; life sciences; earth and space sciences; and engineering, technology, and applications of science.

Topics relevant to the Anthropocene, such as Human Impacts, Human Sustainability, and Climate Change, are evident within the Earth and Space Science standards. The Earth and Space Science standards are broken into three core ideas with three to five subcomponents within each; the second and third cores ideas are relevant to the Anthropocene. The second core idea, *Earth's Systems*, encompasses the processes that drive Earth's conditions, describes Earth's individual systems, and explains how they are interrelated. The third core idea, *Earth and Human Activity*, addresses society's interactions with the planet, explicitly investigating human impacts on Earth's systems and processes. The Earth and Human Activity core idea also includes an entire subsection on global climate change. Crosscutting concepts are "concepts that bridge disciplinary boundaries having explanatory value throughout much of science and engineering" [*NRC*, 2012, p. 83]. The seven crosscutting themes include (1) Patterns, (2) Cause and effect: Mechanism and explanation, (3) Scale, proportion, and quantity, (4) Systems and system models, (5) Energy and matter: Flows, cycles, and conservation, (6) Structure and function, and (7) Stability and change. There are clear connections

---

[1]*Doctoral Candidate, STEM Education Center, Department of Curriculum and Instruction, University of Minnesota, St. Paul, Minneapolis*
[2]*Associate Director and Associate Professor, STEM Education Center, University of Minnesota, Minneapolis, Minnesota*
[3]*Associate Professor, Science Education, Department of Curriculum and Instruction, University of Idaho–Coeur d'Alene*
[4]*Doctoral Student, Department of Curriculum and Instruction, University of Idaho–Coeur d'Alene*

---

*Future Earth—Advancing Civic Understanding of the Anthropocene, Geophysical Monograph 203*, First Edition.
Edited by Diana Dalbotten, Gillian Roehrig, and Patrick Hamilton.
© 2014 American Geophysical Union. Published 2014 by John Wiley & Sons, Inc.

between the Earth's Systems and Earth and Human Activity core ideas and the crosscutting themes.

In terms of classroom practice, the overarching intent of the standards is that teachers should integrate the three dimensions (practices, crosscutting concepts, and core ideas) and students will "actively engage in scientific and engineering practices in order to deepen their understanding of crosscutting concepts and disciplinary core ideas" [*NRC*, 2012, p. 217]. This integration will be challenging for teachers. Indeed, many teachers still do not implement inquiry-based instruction as outlined in the original *National Science Education Standards (NSES)* [*NRC*, 1996; *Johnson*, 2006]. It stands to reason the implementation of the new standards to be developed from the *Frameworks* document [*NRC*, 2012] will require changes in both preservice preparation and professional development for current science teachers.

Furthermore, teachers face additional challenges with the requirement and need to address socially charged or "controversial" topics that surround discussions related to teaching about the Anthropocene. Despite strong arguments for the inclusion of such controversial issues in school curriculum, teachers remain uncomfortable integrating these topics in the classroom. Amid the problems with teaching these controversial subjects, both teachers and students can be left in uncomfortable situations [*Byford et al.*, 2009], become socially divided, or deem material culturally taboo [*McGinnis*, 1992], and thus avoid inclusion in the curricula. Ironically, *Pederson and Totten* [2001] found most teachers report feeling confident in their ability to teach controversial topics and state their importance within the curriculum, however in practice these controversial topics are avoided. If teachers are to successfully implement these current and challenging topics to their students, then professional development is the bridge that can assist teachers in achieving the knowledge, skills, and support they need to effectively teach these new standards to their students.

In this chapter we provide background information for scientists and educators interested in implementing professional development for K–12 teachers as part of their academic work or outreach extensions of their scientific work. First, we provide background on what the literature reports as best practices in science teacher professional development, including evaluation of teacher professional development programs. Second, we share vignettes from four different professional development programs to highlight practical examples of the application of these best practices. Finally, we share a rationale and practical advice for scientists considering working in teacher professional development related to the Anthropocene.

## 3.1. TEACHER PROFESSIONAL DEVELOPMENT

Teachers learn in various ways, most of which is governed by the ease of accessibility to the learning opportunities. Teacher learning can take place in various formal and informal contexts that range from structured, topic-specific seminars, college courses, and in-service workshops to broader science teaching conferences, and one-day district general education programs and seminars. Informal learning can take place in various situations that are usually part of teachers' daily routines, for example, mentoring, reflecting on lessons, and group discussions surrounding curricula and improvement of instruction [*Guskey*, 2000; *Desmione*, 2009]. Traditionally, teacher professional development was approached as an episodic one-shot workshop [*Guskey*, 1986; *Smylie*, 1989; *Stein et al.*, 1999] dominated by a "training" paradigm. Within this traditional paradigm, teacher training was packaged in the form of short-term, discrete workshop sessions that were predominantly held at the district level and treated teachers as passive recipients of knowledge [*Dass and Yager*, 2009].

However, with the advent of reforms focused on inquiry-based instruction [*American Association for the Advancement of Science (AAAS)*, 1993; *NRC*, 1996], teacher professional development practices also moved toward more collaborative and long-term learning experiences. Professional development is viewed as a process that deepens teachers' content knowledge, that leads to changes in classroom practices and ultimately, enhanced student learning [*Desimone et al.*, 2002]. Researchers agree on a set of core features for quality professional development that have been shown to produce successful outcomes for teacher learning [*Darling-Hammond and Richardson*, 2009; *Desimone*, 2009; *Garet et al.*, 2001]. These core features include: content focus, active learning, coherence, duration, collective participation, and program evaluation.

### 3.1.1. Focus on Content

Myriad studies have pointed to a link between a focus on content knowledge in professional development and improvements in teachers' knowledge and skills [*Garet et al.*, 2001; *Smith et al.*, 2007; *Desimone*, 2009]. Content focus in professional development means that activities should be designed to "help teachers better understand both what they teach and how students acquire specific content knowledge and skills" [*Guskey*, 2009, p. 497]. This is particularly critical when the goal is the implementation of inquiry instruction designed to improve students' conceptual understanding [*Fennema et al.*, 1996; *Cohen and Hill*, 1998]. Thus, successful professional development programs that are partnerships between scientists and educators provide a strong focus on both current scientific knowledge and pedagogical transformations of that content to improve teaching and learning [*Roehrig et al.*, 2012b].

### 3.1.2. Active Learning

Professional development activities should provide teachers with opportunities to actively engage in the "meaningful analysis of teaching and learning, for example, by reviewing student work or obtaining feedback on their teaching" [*Desimone et al.*, 2002, p. 83]. It is also critical that professional development activities model the reform-based practices that teachers are expected to implement in their classrooms [*National Science Teacher Association (NSTA)*, 1996]. Thus, rather than listening to a lecture, teachers should experience science through direct immersion in scientific research and long-term experiments [*Loucks-Horsley et al.*, 1998]. These experiences can be integrated into a professional development workshop or a longer term research experience for teachers.

### 3.1.3. Coherence

Professional development will have a greater impact on classroom practices if there is coherence both within and across professional development programs. Coherence calls for teacher learning to be consistent with teachers' knowledge and beliefs, as well as being aligned with school, district, and state policies [*Penuel et al.*, 2007]. It is important that professional development programs use language that parallels national, state, and district standards and policies; any professional development program should directly consider the new *Frameworks* [*NRC*, 2012] and forthcoming Next Generation Science Standards.

### 3.1.4. Duration

Duration considers the total number of contact hours spent with teacher participants during the professional development. Although professional development duration tends to range widely, researchers have reported the number of hours necessary for effective professional development as approximately 30 hours or more [*Desimone*, 2009; *Guskey and Yoon*, 2009]. Equally critical is the distribution of these learning experiences over time; research supports activities that are spread over a semester (or intense summer institutes with follow-up during the semester and even into the academic year) [*Supovitz and Turner*, 2000].

### 3.1.5. Degree of Collective Participation

Participation from the same department, grade, or school enables teachers to work in teams and more effectively integrate their learning from the professional development into their classroom. Additionally, encouraging these teachers to work together provides opportunities for collegial support, professional collaboration, and cohesion of goals and objectives. Collective participation also increases the potential for teacher learning through interaction and discourse provided by opportunities to co-plan and reflect on curriculum, instruction, and assessment [*Borko*, 2004; *Fullan*, 1991; *Guskey*, 1994; *Loucks-Horsley et al.*, 1998].

### 3.1.6. Program Evaluation

Professional development should be a purposeful endeavor, and through the use of a purposeful evaluation, the effectiveness of chosen strategies can be determined. Such evaluation requires the collection and analysis of the five critical levels of information as proposed by *Guskey* [2002] (see Table 3.1).

Information collected at all the five levels is important and provides vital data for evaluating and improving the quality of the professional development program. However, the effectiveness at one level of the program might not be the estimator for the impact at the next level. Time and planned effort, through the five core features described previously, are involved in moving from professional development experiences (Level 1) to improvements in student learning (Level 5). Hence, it is suggested that in planning professional development to improve student learning, the order of these levels must be reversed, starting with the desired student learning outcomes (Level 5) and working backward to determine which professional development experiences will enable participants to succeed at each prior level [*Guskey*, 2001].

## 3.2. EXAMPLES OF TEACHER PROFESSIONAL DEVELOPMENT PROGRAMS

Researchers have frequently debated which types of particular professional development activities or designs are most effective [*Easton*, 2004]. However, research is clear that it is the core features of the professional development, described in the previous section, that determine success rather than the type of activity (face-to-face, online, research experience, etc.). Yet, when planning a professional development, it is necessary to make decisions that relate to the type of activity. In the following section of the chapter, we asked scientists and science educators to share their decisions about engaging in teacher professional development as part of their

**Table 3.1** Critical Levels of Professional Development Evaluation

| Level | Purpose | Method | Key Questions |
| --- | --- | --- | --- |
| 1. Participant Reaction | Evaluate participants' reactions about information and basic needs | Survey Questionnaire | Did they like it? Was your time well spent? Was the presenter knowledgeable? |
| 2. Participant Learning | Examine and evaluate participants' level of attained learning | Content assessments Personal reflection Concept maps Participant portfolios | Did participants acquire intended knowledge and skills? |
| 3. Organizational Support and Learning | Analyze organizational support for skills gained in staff development. | Surveys or interviews with school or district administrators | Was the implementation facilitated and supported? Were sufficient resources made available? |
| 4. Participant Use of New Knowledge and Skills | Evaluate whether participants are using what they learned | Surveys of classroom practices Teacher reflections Direct classroom observations | Are participants implementing their knowledge and skills and to what degree? |
| 5. Student Learning Outcomes | Evaluate the impact on student learning, student performance, or achievement. | Classroom grades Content tests Laboratory reports Direct observation of classrooms | Did students show improvement in academic, behavior, or other areas? |

education and outreach grant requirements and provide illustrative examples of successful professional development programs that reflect different decisions about the structure of professional development. The vignettes illustrate different approaches to teacher professional development and are used to highlight different components of quality professional development. The first vignette describes two face-to-face professional development programs developed to promote climate change education in American Indian Communities. The second vignette, provided by the American Meteorological Society, describes online approaches to teacher professional development. The third vignette describes a professional development approach used at Oregon State University to immerse teachers in field-based research. Finally, the Cooperative Institute for Research in Environmental Sciences (CIRES) program provides suggestions for connecting geoscientists and educators to promote the teaching of Anthropocene-related science.

### 3.2.1. Vignette: Teacher Professional Development for Climate Change Education in American Indian Communities

CYCLES and the Intermountain Climate Education Network (ICE-Net) are two separate teacher professional development efforts whose goals are to promote climate change education in American Indian communities. The CYCLES program works with teachers in American Indian communities in Northern Minnesota and ICE-Net with teachers on American Indian reservations throughout Idaho and Northeastern Washington. Although these are two separate NASA Innovations in Climate Education (NICE) teacher professional development programs they are similar in scope and design.

Both programs include a focus on the science of climate change (content focus), engagement and creations of culturally relevant climate change lessons and activities (active learning), an exploration of personal and societal attitudes and beliefs about climate change (coherence), a series of multiday summer workshops and school-year follow-up activities over the three years of the projects (duration), recruitment and participation of teachers working with American Indian students (collective participation), and comprehensive evaluation programs. Of particular importance is the recognition that most of the professional development participants were the only science or science-related subject teacher at the school or in the district. Thus, building a community of teachers working in tribal communities allowed for the collaborative development of a "teacher community of practice" to discuss and develop ideas for teaching climate with this particular groups of students.

#### 3.2.1.1. Culturally Congruent Climate Change Education

The CYCLES and ICE-Net programs both four- to eight-day resident summer workshops that engage teachers in understanding climate concepts as articulated by the *Climate Literacy: The Essential Principles of*

*Climate Science* [*National Oceanic and Atmospheric Administration (NOAA)*, 2009]. During the workshops, teachers explore climate change science within the local environment, using culturally relevant teaching and place-based pedagogies. All activities for the CYCLES and ICE-Net professional development programs draw on culturally relevant contexts and align with traditional ecological knowledge [see *Roehrig et al.*, 2012a]. For example, in many American Indian cultures, the medicine wheel symbolizes the interconnectedness of earth, air, water, and fire. This is recognized in science through an Earth systems approach based on the interconnectedness of the geosphere, atmosphere, hydrosphere, and biosphere, with the energy flow of these systems derived from the "fire" of the Sun and the interior of the Earth. The medicine wheel components are thus connected to the central concepts (or earth spheres) needed to understand climate change. In organizing these central concepts through a medicine wheel, climate change content can be explored through a culturally relevant lens.

Where the two programs differ is the structure of follow-up sessions and activities implemented by the two programs. The CYCLES program follow-up sessions were composed of five day-long Saturday experiences throughout the school year. The ICE-Net program held monthly "check-in" meetings for program teachers to touch base, ask for ideas from fellow teachers, and assistance from program scientists and experts. Because of the vast distances between ICE-Net teachers in Idaho and East Washington, the monthly check-in meetings were held via distance and virtual means (i.e., computer, compressed video, and telephone). Both programs enact follow-up activities to continue to engage teachers in learning and reflecting on culturally relevant ways their American Indian students can learn about climate change, encouraging them to draw on the local community and environment around them.

### 3.2.1.2. Evaluation of the Climate Change Professional Development Programs

Research related to CYCLES and ICE-Net has primarily focused on levels two and four of the five-level evaluation framework (see Table 3.1) [*Guskey*, 2000]. Evaluation of these professional development programs at level two has considered changes in teachers' attitudes, beliefs, and knowledge of climate change. During the CYCLES and ICE-Net workshops, teachers were asked to confront their attitudes and beliefs about the content and impact of climate change. We have used the "Six Americas" survey [*Leiserowitz et al.*, 2010] to explore changes in teachers' attitudes toward climate change over time. This survey is designed to categorize attitudes related to climate change along a spectrum, *alarmed, concerned, cautious, disengaged, doubtful*, and *dismissive*. People categorized as alarmed and concerned are convinced that climate change is happening and humans are the main cause, but the former are more worried about it. Those cautious and disengaged are not sure that climate change is happening and have not given much thought to it. The cautious can easily change their mind about climate change and are somewhat worried about this issue, whereas the disengaged usually pay little attention to information regarding it. Those who fall into the doubtful group tend to view climate change as a natural change and are not worried about it. The dismissive believe that human-caused climate change is not happening and this issue is not important to them. In addition, we also used the *New Ecological Paradigm* (NEP) *Scale* [*Dunlap and Van Liere*, 1978; *Dunlap et al.*, 2000]. This scale is intended to assess teachers' beliefs about the relationship between humans and Earth. The scale was designed to examine the degree to which people endorse an "ecological" worldview. The instrument contains items such as "*The balance of nature is very delicate and easily upset*" and "*Humans have the right to modify the natural environment to suit their needs.*"

Given the complex and interdisciplinary nature of climate change science, it is necessary to evaluate teachers' knowledge with more sophisticated and holistic measures than just multiple-choice content assessments. Therefore, using tools like Concept mapping [*Gautier et al.*, 2006] and *Photo Elicitation Interviews* (PEI) [*Epstein et al.*, 2006] is recommended. These approaches not only bring forth prior knowledge of the learners but also depict the logical structure of their understanding [*Wandersee*, 1990]. Teachers work in groups to create concept maps progressively through the professional development experience, which are analyzed both quantitatively and qualitatively [*Kern et al.*, in press]. Analysis of concept maps at the end of the first year of the professional development programs showed that most of the teacher knowledge about global climate change was associated to one core concept and most participants were conceptually at a level where they are unable to support a "systems-perspective" or an interconnected "world-view." Duration of professional development for complex earth system science concepts appears to be particularly critical to allow teachers to develop the necessary knowledge.

PEI use images to elicit longer and more comprehensive discussions on the topics that the photos represent and allow the participants to respond to the questions being asked without indecisiveness and intimidation. The images used for the interviews of participants were obtained from the NASA image collection, and the questions associated with these images were aligned with the essential principles of climate science [*NOAA*, 2009]. The richness of the conversation during a PEI, when an image of the impact of global climate change on local weather patterns was shown during the interview is evident in the following quote from a teacher participant:

"I guess with the temperature rising, there's more evaporation. More evaporation, the more water in the atmosphere, and it's got to go somewhere. I think regionally some areas are going to get more rain, but other areas are not going to get rain, depending on weather patterns."

### 3.2.2. Vignette: American Meteorological Society DataStreme Model of In-service K–12 Teacher Professional Development

Founded in 1919, the mission of the American Meteorological Society (AMS) is to advance atmospheric and related sciences, technologies, applications, and services for the benefit of society. The AMS Education Program promotes the teaching of atmospheric, oceanographic, and hydrologic sciences through precollege teacher professional development and instructional resource material development, along with instructional innovation at the introductory college course level. These educational initiatives include activity directed toward greater human resource diversity in the sciences.

#### 3.2.2.1. DataStreme Teacher Professional Development Courses

The AMS Education Program, DataStreme (Atmosphere, Ocean, and Earth's Climate System) offers professional development courses each fall and spring semester for in-service K–12 teachers throughout the United States. The courses are developed by education program scientists in Washington, D.C., and locally administered by trained local implementation teams. Approximately 60 local implementation teams administer DataStreme courses to about 900 teachers each year. These teams typically consist of an AMS-trained master teacher, a scientist (e.g., meteorologist, oceanographer, or climate scientist), and a college instructor with experience in the discipline.

DataStreme is generally administered online with three face-to-face meetings and one-on-one mentoring. These distance-learning courses are offered at the graduate-level, are content rich experiences in which participants use information technology to explore fundamental scientific concepts, interactions among components of the Earth system (i.e., atmosphere, hydrosphere, cryosphere, geosphere, and biosphere), and the human/societal impacts on and response to those interactions (*content focus*). On successful completion of the course, teachers receive three, tuition-free graduate credits from the State University of New York (SUNY) College at Brockport.

DataStreme closely aligns with the *Framework for K–12 Science Education* [*NRC*, 2012] and the Next Generation Science Standards (NGSS) [*NGSS Lead States*, 2013]. Investigating the scientific basis of the workings of Earth's ocean and atmosphere as components of the Earth System follows the crosscutting theme of the *Frameworks* and the NGSS. The roles of water in Earth's surface processes, weather and climate, and global climate change are NGSS Earth and Space Science Disciplinary Core Ideas. DataStreme courses provide process and content knowledge for elementary-, middle-, and high-school–level educators to teach to the new science standards. Teachers then adapt this process and content to their grade levels (*coherence*).

Datastreme courses incorporate hands-on exploration of real world data, including current NOAA and NASA products and services, thereby modeling the practices of science (*active learning*). DataStreme course components include a 15-chapter textbook, investigations manual, and course website. Participating teachers explore 12 principal themes arranged by chapter, each corresponding to one week of the semester-length course. The investigations manual contains twice-weekly learning activities tied directly to each chapter (*duration*).

In addition, AMS education program scientists write weekly current events (Weather, Ocean, Climate) website investigations, using NOAA or NASA data in most cases. Throughout the course, teachers develop a plan of action outlining their subsequent role as peer-trainer in their school districts and communities. Hence, DataStreme teachers become their school's weather, ocean, or climate representative, many times in collaboration with other teacher colleagues who have taken DataStreme (*collective participation*). A recent AMS survey showed that through the peer-training multiplier effect, a DataStreme trained teacher impacts an average of up to 10 additional teachers and more than 350 students within two years of course completion.

#### 3.2.2.2. Recruitment of In-service K–12 Teachers

AMS proactively reaches out to teachers who are members of groups traditionally under-represented in the STEM disciplines or teach in schools with 25 percent or greater minority student enrollments. DataStreme local implementation teams are located in many areas with large minority-student populations including New York City/Long Island, New York; Camden County, New Jersey; Baltimore, Maryland; Prince Georges County, Maryland; Washington, D.C.; Richmond, Virginia; Virginia Beach, Virginia; south Florida; Gary, Indiana; St. Louis, Missouri; Albuquerque, New Mexico; Phoenix, Arizona; Los Angeles, California; and San Bernardino County, California; with local implementation teams in Baltimore, Richmond, Virginia Beach, Albuquerque, and Iowa City led by teachers who are from under-represented groups. Local implementation teams recruit local teachers for DataStreme courses through educational networks, and attendance at district-based and professional science meetings. AMS staff members actively solicit teachers to sign up for the courses to expand the local recruitment, especially with booth presence and presentations at scientific, educational, and professional meetings such as those organized by the National Science Teachers Association, National Earth Science Teachers Association, National Association for Research in Science Teaching, National Marine Educators Association, and National Council for Geographic Education, as well as the American Geophysical Union. With the upcoming implementation of the NGSS, AMS plans to collaborate with state science supervisors to provide teacher training.

*3.2.2.3. Leadership Training and Renewal Experiences for DataStreme Leaders*

With the continuous developments in the environmental and earth sciences, it is important for local implementation team leaders to stay current and learn about new scientific advances. The AMS conducts annual summer training workshops for the team leaders. These are typically held in partnership with a NOAA or NASA facility and major research university, and expose teachers to operational and research activities. In addition to summer training, many team leaders attend the AMS Annual Meeting, where they assist with WeatherFest and present at the Symposium on Education. Approximately 18 teachers per year complete the Project Atmosphere (1991–present) workshop at NOAA's National Weather Service Training Center in Kansas City, Missouri, and 24 teachers annually attend the US Navy-supported Maury Project (1993–present) workshop at the US Naval Academy in Annapolis, Maryland. Trained teachers conduct workshops for their peers and provide the content expertise and leadership skills needed to deliver DataStreme courses.

*3.2.2.4. Evaluation of DataStreme Courses*

DataStreme courses are rigorously assessed by an external evaluator through ongoing review of course materials and methods that aid in targeted revision of AMS tools for measuring teacher effectiveness. AMS partners with local implementation team leaders to monitor STEM learning outcomes to assess the effectiveness of course materials and pedagogical approaches. This is done via pre- and posttests on pedagogy, process, and content, postcourse surveys, postcourse environmental literacy evaluations, and the evaluation of participant plans of action. AMS and the external evaluator conduct follow up interviews with select participants to assess plan of action implementation and classroom impact and survey the remainder online. This approach allows for a qualitative and quantitative appraisal of participant impact on STEM education. The AMS external evaluator also assesses how AMS meets project milestones, including numbers of DataStreme teachers trained each semester, students and other educators impacted, and key course enhancements, including new editions of materials and the use of current geoscience data.

### 3.2.3. Vignette: Teachers as Watershed Researchers Professional Development Project

The *Teachers as Watershed Researchers* (TWR) project was designed as a partnership between the H. J. Andrews Experimental Forest Long-Term Ecological Research (LTER) Program and the Oregon Natural Resources Education Program at Oregon State University to provide high school teachers with the knowledge, tools, and experience to confidently engage their students in authentic watershed-based science inquiry projects (*content focus*). Through a series of three workshops scheduled throughout the year (*duration*), high school teachers worked closely with research scientists, master teachers and science educators at the Andrews Forest LTER site and other watershed field sites. This approach engaged teachers in scientific inquiry (*active learning*) and provided the opportunity for learning both content and pedagogy to transfer that understanding to their own classroom. An emphasis was placed on supporting teachers in designing watershed-based science inquiry projects that connected students to their communities.

More specifically, in this professional development project 15 teachers from throughout Oregon participated in three two-day workshops, one in the summer (Andrews Forest LTER site), another in the fall (master teacher field site on the Mary's River), and the last in the spring (Oregon Hatchery Research Center) (*duration*). Each workshop was structured similarly, with an emphasis placed on one aspect of the science inquiry process, opportunities to practice at least one field investigation and sharing and problem-solving with other participants at each workshop. By the end of the workshop series, participants had practiced all aspects of the scientific process from developing a researchable question through analyzing and interpreting field data. During the two full days at each field site, participants spent most of the time learning from and interacting with the science researchers. The researchers provided content instruction in classroom, laboratory, and field settings, while discussing the research in which they were engaged. Teachers practiced conducting field investigations (*active learning*) similar to what they might do with their students. They also received instruction and practice with analyzing and interpreting the watershed-based data they collected. Thus, TWR provided the teachers with the what (science content), the how (field techniques, data collection and analysis), and the why (why are specific data collected, what do field scientists actually do with the findings) (*coherence*). Master teachers and science educators facilitated the instruction by providing guiding questions, modeling, and sharing of curricular material on how the science in which the teachers were engaging could be transformed into classroom teaching/learning experiences for their high school students. At each workshop, the teachers were engaged in "make and take" of simple field equipment they would make at the workshop, use in the field, and then take back for use in their own classrooms.

Development of a research proposal as a framework for designing a watershed-based research science inquiry project was the guiding principle for this professional development. The research proposal parallels the science inquiry process defined by Oregon Department of Education Science Standards (*coherence*) and provides the structure to guide teachers and students through the development of a relevant watershed-based science inquiry project. The series of workshops led teachers through the process of: identifying community partners (e.g., watershed council coordinators, Oregon Department of Fish and Wildlife staff); identifying a potential watershed-based project in which they could engage their students; developing a researchable question for the watershed project; describing methodology, tools, and resources to answer the question; and planning how they would engage their students in data analysis and interpretation. Teachers practiced, discussed, /r saw examples of each aspect of the research proposal during the workshop, and the science educators and research scientists gave feedback on the research proposals as the teachers developed them. This approach provided a system of modeling, practice, and feedback over the course of a year to give teachers the confidence and capacity to engage students in authentic watershed-based science inquiry projects.

Evaluation of the professional development project was based on tests, surveys, postworkshop teacher reflections and assignments, which included an overview and description of the watershed-based science inquiry project implemented with their students. Not only did the TWR project help teachers to bridge the gap between scientific inquiry as conducted by researchers and scientific inquiry in which high school students can engage, but it also improved teachers' skills and confidence in using constructivist practices in inquiry learning in their classrooms. Teachers created partnerships in the community and with each other that will help them engage their students in watershed-based science inquiry projects over the long term and successfully transferred their learning from the TWR project into implementation of watershed-based science inquiry projects in their local communities. The Oregon Natural Resources Education Program in partnership with the Andrews Forest LTER Program supported three cohorts of middle- and high-school teachers through similar year-long professional development projects and now provides continuing opportunities for these teachers to collaborate with each other and to deepen their skills in engaging their students in field-based research through extended scientist-teacher partnerships such as the Research Experience for Teachers program and other workshops.

### 3.3. CONNECTING GEOSCIENTISTS AND EDUCATORS

In this final section, Dr. Susan Buhr describes the CIRES Outreach and Education Group and their work to engage scientists in the important work of education related to the Anthropocene. She concludes with advice for scientists considering professional development as part of their education and outreach mission.

Support for educators is one of the ways researchers may choose to increase the societal impact of their work [*NASA*, 2010; *National Science Foundation (NSF)*, 2013]. Leaders in science education have identified engagement of scientists in education as having a meaningful impact on science education [*Bybee*, 1998; *Fraknoi*, 2005 *Alberts*, 2009]. The CIRES outreach program was established in 1996 for three purposes: to make a contribution to education for the community, to support CIRES investigators proposing social impact aspects within their grant requests, and to contribute to the teaching mission of the University [*Buhr*, 2002[. Making a broader societal impact with research projects was being emphasized at the NSF, and many at CIRES wondered how to respond to this criterion in a way that would improve and not diffuse their research efforts, or they viewed the new requirement with trepidation because of lack of experience with education and public outreach activities. From a volunteer-run one-day workshop for local teachers, the CIRES Education and Outreach program has grown into a resource for researchers that includes curriculum development and professional development capability, scientific expertise, knowledge of geoscience education needs and priorities, administrative and marketing support, evaluation services, and communications training for scientists. Although they do serve many audiences, the core of their work connects scientists with educators through face-to-face workshops, online courses, digital learning resources, online community building, field courses and more. CIRES Education and Outreach works with classroom educators from elementary through high school, informal educators, and undergraduate faculty.

All of this capacity benefits principal investigators because it provides infrastructure, expertise, a partner with existing reputation and relationships in the educational community, and credibility. Although the importance and application of broader impacts varies by funding program and agency, in a competitive funding environment a well-developed broader impacts component can make the difference between an excellent proposal that receives funding and an excellent proposal that does not. In addition to the pragmatic benefit of submitting more competitive proposals, good broader impacts activities benefit all parties, educators and scientists alike. Benefits to scientists include gains in communication and teaching skills, satisfaction of contributing to society, and fun [*Andrews*, 2005; *Tanner*, 2000]. Partnerships with geoscientists benefit educators by providing teachers with current research knowledge, a better understanding of the nature of science, a colleague with whom to check content knowledge, and access to new resources and learning activities [*Thompson et al.*, 2002, *Moskal et al.*, 2007]. Even short-term interactions in classrooms provide student and teacher benefits in enhanced interest and engagement, new views of scientists, and an understanding of science and its relevance [*Laursen, et al.*, 2007].

CIRES Education and Outreach made some early programmatic decisions designed to increase the impact of their work. They decided to focus on educators and school districts to achieve a multiplier effect, to provide longer-term professional development based on best practices, to evaluate our efforts, and to be guided by the expressed needs of their partners. To provide continuity beyond the usual grant cycle, they typically develop a suite of similar projects with broader impacts funding to achieve efficiency of scale or they use infrastructure built through their own externally funded projects, which then benefits the work funded in partnership with science teams. Educator-focused projects usually include professional development, curriculum development, evaluation, and possibly a direct component for students. The projects illustrated here are just a few of the past and current professional development projects within the CIRES Education and Outreach portfolio.

### 3.3.1. Climate and Oceans Teacher Professional Development: Connecting the Global Oceans to Inland Audiences

This project combined CIRES Education and Outreach external funding with funding from several NSF research grant broader impacts activities to support an annual summer teacher workshop series over a four-year period, a teacher exchange, and virtual workshop participation during the academic year. Legacy products include videos of scientist presentations, workshop resources such as lesson plans, and a publication [*Smith et al.*, 2013]. Roles for scientists included participating in preparation for their role, giving presentations, and helping teachers with hands-on workshop activities.

#### 3.3.1.1. Weather and Water After-School Student-Scientist Project: Waterspotters

In partnership with the local Math, Engineering and Science Achievement (MESA) program, an enrichment program for students under-represented in STEM disciplines, student-scientists track weather and precipitation data, learn about weather and water through hands-on activities, and collect precipitation samples for isotope analysis by the research team to understand the Colorado Front Range water budget. MESA educators and CIRES Education and Outreach staff co-developed the curriculum with review and input from the science team. After review of the curriculum, piloting MESA teachers will participate in workshops to learn about the topics and the curriculum. Early career scientists (students and postdoctoral researchers) install and support weather stations at the participating middle schools, answer questions, visit sites, and participate in workshops. As weather stations are installed, the data will be available online.

#### 3.3.1.2. Sun-Earth Connections Education and outreach: Extreme Ultraviolet Variability Experiment Geomagnetism in the MESA Classroom

Through several supplements to NASA satellite and science projects, Sun-Earth connection broader impact activities have included teacher and MESA advisor workshops, a journalist workshop, a course for under-represented high school students, digital resources, videos, and MESA after school kits. Scientists review materials, initiate lesson ideas, provide data and imagery, appear in videos, and visit the course and after-school sites. Legacy products include videos, new learning resources, and workshop materials.

#### 3.3.1.3. Authentic Scientific Inquiry: Scientific Inquiry on the Tibetan Plateau: Upward and Outward

In partnership with a research team studying uplift of the Tibetan Plateau, CIRES Education and Outreach developed a video about the nature of science as it is practiced in geoscience. Along with the video, CIRES Education and Outreach conducted workshops for teachers about the nature of geoscience and provided classroom resources and background information about how students perceive science and scientists. Through evaluation, the staff found that the film was successful in challenging stereotypes and showing science as a human process of building knowledge and working together [*Laursen and Brickley*, 2011a, 2011b].

### 3.3.2. Strategies for Success

In the following section the CIRES team identifies some strategies for scientists interested in engaging in professional development as part of their education and outreach work.

**Find partners who use best practices.** Every field has research-based standards for best practices; you should choose your collaborators based on their demonstrated expertise. Any work with educators is most effective when grounded in the best practices in the field. Ask your potential partners how they use the research base in education and outreach to inform their projects.

**Think of your outreach partner as you do your other collaborators.** The best broader impacts projects are developed like the best research proposals. This includes allowing sufficient time to define the project and bring in any other partners, establishing specific roles and responsibilities, increasing cohesion between the project components, and including sufficient resources for the scope of the effort. Once funded, include your outreach partner in your project communications and activities. If your institute or department does not include outreach professionals, ask your University Continuing Education unit or College of Education for potential partners.

**Meet the needs of schools.** The forthcoming Next Generation Science Standards [*NRC*, 2012] place Earth and environmental sciences on an equal footing with physics, chemistry, and biology, with a new fundamental emphasis on the practices of science. This is a watershed moment in the history of geoscience education, one that could fundamentally change the quality and quantity of student learning in our discipline. However, states and school districts will need help from scientists and geoscience education professionals to develop curriculum and prepare teachers so as to meet the vision. Broader impacts activities within research may help by providing the current state of knowledge on publicly controversial topics such as climate and energy science and preparing more teachers to instruct on foundational geoscience concepts, to understand the practices of geoscientists, and to benefit from connections to working scientists.

**Choose roles you will enjoy.** Many roles are available to scientists, which reflect the researcher's time, expertise, interests, and personality. Within CIRES Education and Outreach projects scientists appear in videos, visit classrooms, provide e-mail support, review resources, present at workshops or webinars, and provide data, images and animations. Having education partners allows scientists to bring their content knowledge expertise to the forefront and not have to learn how to develop and implement full professional development offerings.

**Develop two-way relationships.** Scientists and teachers benefit most from outreach when real relationships develop, when mutual learning happens, and when both kinds of expertise, science and education, are valued and explicitly recognized [*Tomanek et al.*, 2005]. Participate in preparation sessions designed to increase the effectiveness of your contribution, spend time with participants in workshops, bring your students and stay for lunch. All of these things signal to educators that you value your time with them and their expertise.

**Include assessment and evaluation.** Professional outreach providers use social science research methodologies to understand the extent to which education and public outreach activities are meeting the intended goals. Include resources and expertise within your broader impacts activities to identify ways the project is meeting goals, establish learning gains for educators and students, determine ways in which the activity can be improved, and to identify the outcomes and value of the project.

Researchers approach broader impacts with greater and lesser enthusiasm and experience, but realize benefits through participation. Finding a partner who can make this aspect of your research work easy and effective is vital, as is including that partner in your project as a team member on receiving funding. Because of the advent of the NGSS, this is poised to be a watershed moment in geoscience education history in the United States. The scientist's contribution of time, data, access to working scientists, and current research helps that moment to become a legacy of better geosciences preparation for citizens and the next generation of scientists. There are many possible ways to work with educators, with roles for scientists with different predispositions and skill sets. Professional broader impact projects include evaluation using established methodologies. More and more geoscientists conceive of their broader impacts work as part of the regular life of the professional scientist, to the benefit of students and teachers, society and the scientists themselves.

### 3.4. ACKNOWLEDGMENTS

Cycles and ICE-NET are NASA Innovations in Climate Education programs funded under Grant Numbers NNXlOAT53A and NNY10AT77A.

DataStreme contributors: James A. Brey, Ph.D., Director, AMS Education Program; Elizabeth W. Mills, MS, Associate Director, AMS Education Program.

Teachers as Watershed Researchers contributors: Kari O'Connell, Oregon State University and Tisha Morrell, Portland State University.

## REFERENCES

Achieve, Inc. (2013), Conceptual shifts, in Next Generation Science Standards, retrieved November 8, 2013, from http://www.nextgenscience.org/next-generation-science-standards.

Alberts, B. M. (2009), Redefining science education, *Science*, 323, 437.

American Association for the Advancement of Science (1993), *Benchmarks for Science Literacy*, Oxford University Press, New York.

Andrews, E., D. Hanley, J. Hovermill, A. Weaver, and G. Melton (2005), Scientists and public outreach: Participation, motivations, and impediments, *JGE*, 53(3), 281–293.

Borko, H. (2004), Professional development and teacher learning: Mapping the terrain, *Educ. Researcher*, 33(8), 3–15.

Buhr, S. M. (2002), CIRES, 1967–2002: Cooperative Institute for Research in Environmental Sciences, in C. Kisslinger (Ed.), *Pioneering a Successful Partnership*, retrieved November 1, 2013, from http://cires.colorado.edu/about/history/CIRES1967-2002.pdf.

Bybee, R. (1998), Improving precollege science education, the role of scientists and engineers, *J. Coll. Sci. Teaching*, 27, 324–328.

Bybee, R. W. and S. Loucks-Horsley (2001), National Science Education Standards as a catalyst for change: The essential role of professional development, in R. Bowers, P. Bowers, and J. Arlington (Eds.), *Professional Development Planning and Design* (pp. 1–12), National Science Teachers Association, Arlington, VA.

Byford, J., S. Lennon, and W. S. Russell III (2009), Teaching controversial issues in the social studies: A research study of high school teachers, *Clearing House*, 82(4), 165–170.

Congress Joint Economic Committee (2012), STEM Education: Preparing for the jobs of the future, retrieved November 1, 2013, from http://www.jec.senate.gov/public/index.cfm?a=Files.Serve&File_id=6aaa7e1f-9586-47be-82e7-326f47658320.

Cohen, D. K., and H. C. Hill (1998), *Instructional Policy and Classroom Performance: The Mathematics Reform in California, Consortium for Policy Research in Education*, University of Pennsylvania (CPRE RR-39), Philadelphia, PA.

Darling-Hammond, L., and N. Richardson (2009). Teacher learning: What matters? *Educational Leadership*, 66(5), 46–53.

Dass, P. M., and R. E. Yager (2009), Professional development of science teachers: History of reform and contributions of the STS-based Iowa Chautauqua Program, *Sci. Ed. Rev.*, 8(3), 99–111.

Desimone, L., A. C. Porter, M. Garet, K. S. Yoon, and B. Birman (2002), Does professional development change teachers' instruction? Results from a three-year study, *Ed. Eval. and Policy Anal.*, 24(2), 81–112.

Desimone, L. M. (2009), Improving impact studies of teachers' professional development: Toward better conceptualizations and measures, *Ed. Researcher*, 38(3), 181–199.

Dunlap, R. E. and K. D. Van Liere (1978), The new environmental paradigm: A proposed measuring instrument and preliminary results, *J. Environ. Ed.*, 9, 10–19.

Dunlap, R. E., K. D. Van Liere, A. G. Mertig, and R. E. Jones (2000), Measuring endorsement of the New Ecological Paradigm: A revised NEP Scale, *J. Soc. Issues*, 56(3), 425–442.

Easton, L. B. (2004), *Powerful Designs for Professional Learning, National Staff Development*, Council Press, Oxford, OH.

Epstein, I., B. Stevens, P. McKeever, and S. Baruchel (2006), Photo elicitation interview (PEI): Using photos to elicit children's perspectives. *Int. J. Qual. Method*, 5(3), retrieved November 1, 2013, from http://ejournals.library.ualberta.ca/index.php/IJQM/article/view/4366/3496.

Fennema, E., T. P. Carpenter, M. L. Franke, L. Levi, V. R. Jacobs, and S. B. Empson (1996), A longitudinal study of learning to use children's thinking in mathematics instruction. *J. Res. Math. Ed.*, 27(4), 403–434.

Fullan, M. G. (1991), The meaning of educational change. In M. G. Fullan, *The new meaning of educational change* (pp. 30–46), Teachers College Press, New York.

Fraknoi, A. (2005), Steps and missteps toward an emerging profession, *Mercury*, (September4–October), 19–25.

Garet, M., A. Porter, L. Desimone, B. Birman, and K. S. Yoon (2001), What makes professional development effective? Results from a national sample of teachers. *Am. Ed. Res. J.*, 38(4), 915–945.

Gautier, C., K. Deutsch, and S. Rebich (2006), Misconceptions about the greenhouse effect, *JGE*, 54(3), 386–395.

Guskey, T. R. (1986), Staff development and the process of teacher change, *Ed. Researcher*, 15(5), 5–12.

Guskey, T. R. (1994), Results-oriented professional development: In search of an optimal mix of effective practices. *J. Staff Dev.*, 15(4), 42–50.

Guskey, T. R. (2000), Professional development and teacher change. Teachers and teaching: *Theory and practice*, 8, 381–391.

Guskey, T. R. (2001), The backward approach, *J. Staff Dev.*, 22(3), 60.

Guskey, T. R. (2002), Does it make a difference? Evaluating professional development. *Ed. Leadership*, 59(6), 45–51.

Guskey, T. R. (2009). Closing the knowledge gap on effective professional development. *Educational Horizons*, 87(4), 224–233.

Guskey, T. R., and K. S. Yoon (2009), What works in professional development? *Phi Delta Kappan*, 90(7), 495–500.

Johnson, C. C. (2006), Effective professional development and change in practice: Barriers science teachers encounter and implications for reform, *School Sci. Math.*, 106(3), 150.

Kern, A. L., G. H. Roehrig, D. Bhattacharya, J. Wang, F. Finley, B. Reynolds, B., et al. (in press). Drawing on Place and Culture in a climate change education in Native Communities. In Mueller, M. and Tippins, D. (Eds.) *EcoJustice, Citizen Science and Youth Activism: Situated Tensions for Science Education*.

Laursen, S., C. Liston, H. Thiry, and J. Graf (2007), What good is a scientist in the classroom? Participant outcomes and program design features for a short-duration science outreach intervention in K-12 classrooms, *CBE-Life Sci. Ed.*, 6, 49–64.

Laursen, S. L., and A. L. Brickley (2011a), Focusing the camera lens on the nature of science: Evidence for the effectiveness of documentary film as a Broader Impacts strategy. *JGE*, 59, 126–138.

Laursen, S. L., and A. L. Brickley (2011b), A scientist has many things to do: E/O strategies that focus on the processes of science, in J. B. Jensen, J. G. Manning, and M. Gibbs (Eds.), *Earth and Space Science: Making Connections in Education and Public Outreach*, ASP Conference Series 443, 116–124.

Leiserowitz, A., E. Maibach, C. Roser-Renouf, and N. Smith (2011), *Global Warming's Six Americas, May 2011*, Yale Project on Climate Change Communication, Yale University and George Mason University, New Haven, CT.

Loucks-Horsley, S., P. W. Hewson, N. Love, and K. E. Stiles (1998), *Designing Professional Development for Teachers of Science and Mathematics*, Corwin Press, Thousand Oaks, CA.

McGinnis, R. (1992), The taboo and "Noa" of teaching science-technology-society (STS): A constructivist approach to understanding the rules of conduct teachers live by, paper presented at the Annual Meeting of the Southeastern Association for the Education of Teachers of Science, Wakulla Springs, FL, February 14–15.

Moskal, B. M., C. Skokan, L. Kosbar, A. Dean, C. Westland, H. Barker, et al. (2007), K–12 Outreach: Identifying the broader impacts of four outreach projects, *J. Engrg. Ed.*, 96 (2), 173–189.

NASA (2010), Explanatory Guide to the NASA Science Mission Directorate Education and Public Outreach Evaluation Factors, version 3.1, retrieved November 1, 2013, from http://science1.nasa.gov/media/medialibrary/2010/12/01/SMD_MissionGuideV3_1_508.pdf.

National Academy of Sciences, National Academy of Engineering, and Institute of Medicine of the National Academies (2007), *Rising Above the Gathering Storm: Energizing and Employing America for a Brighter Economic Future*, National Academies Press, Washington, DC.

National Oceanic and Atmospheric Administration (2009), *Climate Literacy: The Essential Principles and Fundamental Concepts* (2nd ed.,), retrieved November 1, 2013, from http://oceanservice.noaa.gov/education/literacy/climate_literacy.pdft.

National Research Council (1996), *National Science Education Standards*, National Academy Press, Washington, DC.

National Research Council (2012), *National Science Education Standards*, National Academy Press, Washington, DC.

National Science Foundation (2013), Grant Proposal Guide: Proposal Preparation Instructions, retrieved November 1, 2013, from http://www.nsf.gov/pubs/policydocs/pappguide/nsf13001/gpg_3.jsp.

National Science Teachers Association (1996), *NSTA Position Statement: Professional Development in Science Education*. Retrieved November 8, 2013, from http://www.nsta.org/about/positions/profdev.aspx.

NGSS Lead States (2013), *Next Generation Science Standards: For States, By States*, The National Academies Press, Washington, D.C..

Pedersen J. E. and S. Totten (2001), Beliefs of science teachers towards the teaching of science/technological/social issues: Are we addressing national standards? *Bull. Sci. Technol. Soc.*, 21(5), 376–393.

Penuel, W. R., B. J. Fishman, R. Yamaguchi, L. Gallagher, and P. Lawrence (2007), What makes professional development effective? Strategies that foster curriculum implementation, *Am. Ed. Res. J.*, 44(921), 921–958.

President's Council of Advisors on Science and Technology (2010), *Federal Science, Technology, Engineering, and Mathematics (STEM) Education: Five-Year Strategic Plan*, retrieved November 1, 2013, from http:///www.whitehouse.gov/ostp/pcast.

Roehrig, G. H., K. M. Campbell, D. Dalbotten and K. Varma (2012a), CYCLES: A Culturally-relevant approach to climate change education in Native communities, *J. Curr. Instr.*, 6, 73–89.

Roehrig, G. H., M. Michlin, L. Schmitt, C. MacNabb, and J. M. Dubinsky (2012b), Teaching neuroscience to science teachers: Facilitating the translation of inquiry-based teaching instruction to the classroom. *CBE- Life Sci. Educ.*, 11(4), 413–424.

Smith, T. M., L. M. Desimone, T. Zeidner, A. C. Dunn, M. Bhatt, and N. Rumyantseva (2007), Inquiry-oriented instruction in science: Who teaches that way? *Ed. Eval. Policy Anal.*, 9(29), 169–199.

Smith, L. K., M. Barber, L. Duguay and L. Whitley (2013), Using the Ocean Literacy Principles to connect inland audiences to the global oceans, *Current*, 28, 2–7.

Smylie, M. A. (1989), Teachers' views of the effectiveness of sources of learning to teach, *Elem. School J.*, 89(5), 543–558.

Stein, M. K., M. S. Smith, and E. Silver (1999), The development of professional developers: Learning to assist teachers in new setting in new ways. *Harvard Ed. Rev.*, 69 (3) 237–269.

Supovitz, J. A., and H. M. Turner (2000), The effects of professional development on science teaching practices and classroom culture, *J. Res. Sci. Teaching*, 37(9), 963–980.

Tanner, K. (2000), Evaluation of Scientist-Teacher Partnerships: Benefits to Scientist Participants, presented at National Association for Research in Science Teaching Conference, April 29, 2000, New Orleans, LA.

Thompson, S. L., V. Metzgar, A. Collins, M. D. Joeston, and V. Shepherd (2002), Examining the influence of a graduate teaching fellows program on teachers in grades 7–12, Proceedings Paper presented at the Annual International Conference of the Association for the Education of Teachers in Science, Charlotte, NC, January 9–12.

Tomanek, D., N. Moreno, S. C. R. Elgin, S. Flowers, V. May, E. Dolan, et al. (2005), Points of view: Effective partnerships between K–12 and higher education. *Cell Biol. Ed.*, 4, 28–37.

Wandersee, J. (1990), Concept mapping and the cartography of cognition, *J. Res. in Sci. Teaching*, 27(10), 923–936.

# 4

## Climate Literacy and Scientific Reasoning

### Shiyu Liu[1], Keisha Varma[2], and Gillian Roehrig[3]

Global climate change has been widely discussed in the public media for the past decades. However, evidence shows that a considerable percentage of the public still dismisses its seriousness [e.g., *Leiserowitz et al.*, 2011]. To increase and improve the public's awareness about the urgency of global climate change issues, science and environmental educators have considered it a main priority to enhance climate literacy. In general, to be literate in climate science, one needs to "understand the essential principles of Earth's climate system; know how to assess scientifically credible information about climate; communicate about climate and climate change in a meaningful way; and be able to make informed and responsible decisions with regard to actions that may affect climate" [*National Oceanic and Atmospheric Administration (NOAA)*, 2009, p. 4].

There have been ongoing debates about what may contribute to the development of climate literacy. Many scholars suggest that to enhance climate literacy in the general public, climate science should be made more accessible and approachable. Those adopting a knowledge deficit model [*Bauer, et al.*, 2007] assume that increased exposure to scientific information about global climate change will lead to increased understanding. However, inconsistent findings from previous research in environmental education show that acquiring more scientific knowledge about global climate change does not necessarily result in higher awareness of this issue or more positive environmental decision making [*Kellstedt et al.*, 2008].

Thus, recent research efforts have focused on exploring factors that may confound the relationship between climate literacy and awareness of global climate change.

Factors, such as pre-existing values and ideological orientations, may act as a "perceptual screen" for the knowledge individuals take in and thus impact their perspectives toward global climate change [e.g., *Maibach et al.*, 2008]. Although such findings enable a more in-depth understanding of influential factors for the development of climate literacy, it also leads to more uncertainty about what effective approaches can be taken to promote climate change education. To answer this question, there is an immediate need to investigate the mechanisms that underlie the interconnections among these factors.

In this chapter, we introduce an important cognitive construct, *scientific reasoning*, to propose a new perspective toward the development of climate literacy. As an underlying mechanism for our mental activities, scientific reasoning relates to different aspects involved in the understanding of climate science. An in-depth investigation of scientific reasoning and its relationship with climate change education can provide an overarching framework for the ongoing efforts in enhancing climate literacy. We will first present a brief review on the nature of scientific reasoning, and then discuss the importance of promoting scientific reasoning in climate change education. Finally, this chapter will conclude with some suggestions on how to incorporate scientific reasoning into the endeavor of enhancing climate literacy.

### 4.1. WHAT IS SCIENTIFIC REASONING?

To further our discussions in this chapter, we will first answer the question, *what is scientific reasoning?*

Broadly speaking, reasoning is a process of drawing conclusions from principles and evidence that are already known to infer new conclusions [*Wason and Johnson-Laird*, 1972]. Reasoning consists of mental processes involved in

---

[1] Department of Educational Psychology, University of Minnesota, Minneapolis, Minnesota
[2] Assistant Professor, Department of Educational Psychology, University of Minnesota, Minneapolis
[3] Associate Director and Associate Professor, STEM Education Center, University of Minnesota, Minneapolis, Minnesota

*Future Earth—Advancing Civic Understanding of the Anthropocene, Geophysical Monograph 203*, First Edition.
Edited by Diana Dalbotten, Gillian Roehrig, and Patrick Hamilton.
© 2014 American Geophysical Union. Published 2014 by John Wiley & Sons, Inc.

generating and evaluating logical arguments [*Anderson*, 1990] and requires skills such as clarification, basis, inference, and evaluation [*Ennis*, 1987]. *Inhelder and Piaget* [1958] defined reasoning as the capability to undertake a set of logico-mathematical operations, such as logical reasoning, probabilistic thinking, and manipulating abstract variables.

Later, scholars argued that *Inhelder and Piaget's* [1958] view of scientific reasoning as a system of "logic-driven" operations undervalued the role of knowledge and contexts in reasoning [e.g., *Dunbar*, 2001]. More recent research has thus emphasized exploring scientific reasoning in domain-specific scenarios [e.g., *Dolan and Grady*, 2010]. As *Schauble* [1996] suggested, "appropriate knowledge supports the selection of appropriate experimentation strategies, and the systematic and valid experimentation strategies support the development of more accurate and complete knowledge" (p. 118). Indeed, scientific reasoning serves as an important strategy for learners to "make associations among new mental sets with already-existed hierarchical structure-based memory" [*She and Liao*, 2010, p. 91].

In the context of science education, scientific reasoning has been widely defined as an explanatory mechanism that individuals apply when trying to understand available information [*Schauble*, 1996]. It is considered to involve a set of critical skills required in scientific inquiry and necessary to engage in conceptual changes, such as developing testable hypotheses, generating experimental designs, controlling variables, coordinating theory and evidence, and responding to anomalous evidence [*Lawson*, 2005].

Within the scope of this definition, it is natural to consider scientific reasoning as a "private enterprise" for scientists. Admittedly, the reasoning skills mentioned previously are essential in scientists' work. However, scientific reasoning entails a broader notion beyond the formal, rigorous reasoning required in laboratory inquiry. Scientists also employ informal reasoning to gain insights into the natural world, as does the general public in their everyday thinking processes [*Sadler*, 2009]. Likewise, *Tweney* [1991] also suggested that, although the results of science may be presented in the language of formal reasoning and logic, the results themselves originate through informal reasoning. Here, informal reasoning is considered as the thinking processes individuals engage in when faced with ill-structured controversial problems, for which there is limited access to necessary information and no indisputable solutions [*Means and Voss*, 1996]. A question such as, "What do you think about _____?" can be an ill-structured problem. For example, when people are asked "*What do you think about global climate change*", the responses to be expected are open ended, and there is no one correct answer to this question. Individuals can employ evidence from their daily lives and draw on their existing beliefs and theories in their answers, and the thinking process involved is informal reasoning.

*Sadler and Donnelly* [2006] proposed the concept of *socioscientific reasoning* to expand the discussions on informal reasoning. Reasoning about socioscientific issues requires individuals to move beyond mere logical thinking and begin to incorporate other forms of reasoning that will allow them to consider complex issues such as global climate change. This type of reasoning involves conceptualizing the inherent complexity of the issues, precluding simple, linear cause-effect reasoning, as well as constructing and evaluating arguments constructed from multiple perspectives in contentious dilemmas [*Kuhn*, 1993; *Sadler*, 2009]. Based on this perspective, *Sadler and colleagues* [2007] characterized four general dimensions in socioscientific reasoning, which are:

1. recognizing the inherent complexity of socioscientific issues;
2. examining these issues from multiple perspectives;
3. appreciating that these issues are subject to ongoing inquiry; and
4. exhibiting skepticism when presented potentially biased information.

Socioscientific reasoning expands the notion of scientific reasoning and initiates further discussions regarding the influence of scientific reasoning on our everyday lives. *Kuhn* [1993] cited a quote from Albert Einstein [1954, p. 290] to illustrate the link between scientific reasoning and everyday thinking:

> The whole of science is nothing more than a refinement of everyday thinking. It is for this reason that the critical thinking of the physicist cannot possibly be restricted to the examination of the concepts of his own specific field. He cannot proceed without considering critically a much more difficult problem, the problem of analyzing the nature of everyday thinking.

Therefore, overall, in this chapter, we define scientific reasoning from two perspectives. On the one hand, we consider that scientific reasoning consists of formal reasoning skills involved in laboratory scientific inquiry and is closely related to logical, mathematical thinking. On the other hand, scientific reasoning entails skills that may be involved to understand and evaluate ill-structured social issues, as well as appreciating scientific findings as reported in the popular media [*Giere et al.*, 2006]. Together, these perspectives can account for the ways individuals in our society encounter and understand complex, socioscientific issues such as global climate change.

## 4.2. WHY IS SCIENTIFIC REASONING IMPORTANT IN CLIMATE CHANGE EDUCATION?

The publication of the *Second Assessment Report of the Intergovernmental Panel on Climate Change* (IPCC) signaled a broad agreement among climate scientists that anthropogenic climate change is underway [*Houghton et al.*, 1996]. IPCC's Fourth Assessment Report further stated that "warming of the climate system is unequivocal" and "most of the observed increase in global average temperatures since the mid-20th century is very likely due to the observed increase in anthropogenic greenhouse gas concentrations" [IPCC, 2007, pp. 2, 5]. Despite the strong scientific consensus on the detection, attribution, and risks, public opinion surveys have shown a decreasing level of awareness among Americans about the threat of global climate change [*Gallup*, 2010].

One reason for the diminished public concern in the face of strong scientific consensus is that climate science is complex. To appreciate the complexity of the climatic system, it is important to understand the interconnections among the atmosphere, hydrosphere, cryosphere, lithosphere, and biosphere [*Lucarini*, 2004]. Full comprehension of these interconnections requires a systematic, scientific thinking approach [*NOAA*, 2009]. Besides, many factors outside the realm of science, such as exceptionally diversified media environment, the framing of media messages, individuals' deeply held values and beliefs, and cognitive flexibility, can also act as a mental filter for information about global climate change and result in various levels of climate literacy [e.g., *Maibach et al.*, 2008]. Therefore, even among those who are well educated in science and recognize the growing scientific evidence of global climate change, it is inevitable that many will overcredit their personal experiences and underweight scientific studies.

For example, although the scientific community has made great efforts to clarify that any one extreme event or extreme weather season could not be directly attributed to global climate change, the public still attributes strong associations between the two [*Bostrom and Lashof*, 2007]. Faced with tragedies such as Hurricanes Katrina and Sandy, many people quickly draw causal conclusions that global climate change is the reason for it. They tend to seek reasons to support and confirm their opinions instead of evaluating sufficient alternative explanations [*Cialdini*, 2009]. Being overconfident about their own knowledge, people claim causal relationships regardless of the lack of sufficient evidence. Such reasoning bias abounds in not only discussions of extreme weather events but also other climate change issues [*Bostrom and Lashof*, 2007], leading to greater potential for misconceptions and confusions. Therefore, there needs to be an immediate emphasis on the development of appropriate reasoning skills to enhance climate literacy and avoid additional misconceptions.

Educational and policy documents have included scientific reasoning as a prominent component in science and environmental education. The *Framework for 21st Century Skills* [*Partnership for 21st Century Skills*, 2006] acknowledges reasoning as a crucial component in the development of cognitive skills and stresses the importance of appropriately using various types of reasoning for controversial issues such as global climate change. Effective application of scientific reasoning will contribute to the effectiveness and reliability of climate change communication: the reasoning processes help individuals to devise and evaluate arguments intended to persuade others as well as gaining an understanding of others' arguments to generate a rebuttal [*Mercier and Sperber*, 2011]. In other words, scientific reasoning increases the information that can be shared in society both in quantity and epistemic quality.

More importantly, *A Framework for K–12 Science Education* [*National Research Council*, 2012] places heavy emphasis on developing reasoning and argumentation skills in climate change education. It is suggested that students develop reasoning skills that can help them evaluate the causes and effects of global climate change. Students are more likely to reevaluate their prior knowledge and beliefs in depth during scientific reasoning. Appropriate application of reasoning facilitates the revision of mental models about global climate change and helps them identify and correct their misconceptions. The standards also emphasize the importance of engaging students in evidence-based arguments about human impacts on the earth climate system and encouraging them to propose, test, and modify possible solutions to current climate issues. In building support for their arguments through reasoning with evidence, as well as designing and justifying solutions for reducing negative human impacts on the environment, students learn to value scientific findings more and become more aware of the urgency of climate issues.

Although the importance of scientific reasoning in climate change education has been established, there have not been many detailed discussions on effective approaches to promote scientific reasoning. In the next section, we will suggest some potential ways that may help scientists and educators to incorporate scientific reasoning into climate change education.

## 4.3. HOW CAN APPLYING SCIENTIFIC REASONING ENHANCE CLIMATE LITERACY?

Because of the complexity of climate science, complex reasoning processes are involved when one attempts to understand climate modeling, interpret scientific findings, support or adapt their existing theories with scientific evidence, and make solid arguments about climate issues. Here, we will focus on three aspects of scientific reasoning that are most essential for enhancing climate literacy: reasoning with analogies, evidence-based argumentation, and epistemological reasoning.

### 4.3.1. Reasoning with Analogies

Analogies are "a key component of human mental life" [*Dunbar*, 2001, p. 315]. They are comparisons made between a familiar domain of knowledge and another less familiar one [*Orgill and Bodner*, 2005]. These comparisons convey systematic mapping of relational structures shared by the two domains, despite arbitrary degrees of differences in the objects that make up the domains [*Chan et al.*, 2012]. For example, astronomer Johannes Kepler likened the sun and planet to two magnets that approach or repel each other depending on which poles are proximate. This analogy is based on the idea that the causal relations between two magnets are the same as those between the sun and planet; it is not according to the physical resemblance between a magnet and the sun [*Gentner and Markman*, 1997].

Analogies are a key component of all aspects of scientific reasoning and have pivotal impact on conceptual learning [*Dunbar*, 1997]. They have played a critical role in promoting scientific progress, and are prevalent in our everyday lives [*Harrison*, 2008]. Analogies can serve as a bridge between prior knowledge and new information: they make new ideas intelligible and initially plausible by relating them to already familiar information [*Chan et al.*, 2012]. Reasoning with analogies provides concrete reference that can be used to visualize abstract concepts, orders of magnitude, or unobservable phenomena [*Orgill and Bodner*, 2005]. Moreover, using analogies can also play a motivational role in meaningful learning [*Glynn and Takahashi*, 1998]. Studies with students show the use of analogies can result in better engagement and interactions in the classroom [*Lemke*, 1990]. The familiar language of an analogy can give us a way to express our understanding and interact with the complex scientific concepts that may otherwise make us too anxious or uncomfortable to explore.

In climate change education, reasoning with analogies is especially beneficial in promoting the development of climate literacy. Global climate change issues are inherently abstract and the nature of global climate change is virtually impossible to detect and experience. Thus, acquiring knowledge about global climate change is particularly difficult and most people only understand global climate change to the extent that it impacts their everyday life [*Kearney*, 1994]. As we are bombarded by an almost infinite amount of information every day, we cannot possibly process and store all the information we encounter about global climate change. Therefore, to effectively communicate about climate issues, techniques that are compatible with human thinking and reasoning processes should be emphasized and employed, and reasoning with analogies is one of them. When appropriate analogies are used to distribute information about global climate change, people are more likely to retain the information in their memory and be willing to apply it when making environmental decisions [*Glynn and Takahashi*, 1998].

The "blanket analogy," for instance, has often been used as an alternative explanation for the greenhouse effect. As *Taylor* [1991] explained, basically, the greenhouse gases in the atmosphere lying over the surface of the Earth act like a blanket. A thickening "blanket" of carbon dioxide traps more heat in the atmosphere, and the blanket is effective against outgoing energy, producing a warming of the Earth. Research has shown that when people are presented with this analogy, their responses and understanding of greenhouse effect improved markedly [*Bostrom and Lashof*, 2007]. Researchers have also used other analogies to explain the scientific processes underlying global climate change. For example, *Moxnes and Saysel* [2008] explored the effectiveness of explaining the process of carbon dioxide accumulation with air mattress and balloon analogies. Such analogies make the scientific information more approachable and interesting, and people can relate to them without the anxiety that may be experienced when it comes to complex science.

Moreover, analogies can also help people better understand the meaning and scope of global climate change and choose appropriate actions [*Halford and Sheehan*, 1991]. Because analogies provide easy-to-access support for individual thinking, when scientists adopt familiar analogies to explain global climate change issues, people may be more motivated to explore the science behind their personal experiences. They also will be more likely to take sound environmental actions when given analogies, such as the one of buying insurance that *Schneider* [2007] employed to argue for acting on global climate change:

> A continuation of "business as usual" raises a serious concern from the risk-management point of view, given that the likelihood of warming beyond a few degrees before the end of this century (and its associated impacts) is a better than even bet. Few security agencies, businesses or health establishments would accept such high odds of potentially dangerous outcomes without implementing hedging strategies to protect themselves, societies and Nature from the risks-of climate change in our case. This is just a planetary scale extension of the risk-averse principles that lead to investments in insurance, deterrence, precautionary health services and business strategies to minimize downside risks of uncertainty. … We buy fire insurance for our house and health insurance for our bodies. We need planetary sustainability insurance.

Despite the advantages of analogies, arguments exist regarding the drawbacks of reasoning with analogies. *Kearney* [1994] discussed that analogies may hamper the understanding of global climate change if not precisely analogous to the scientific processes involved. Although the blanket analogy mentioned previously makes it easier to understand the greenhouse effect, there have been a lot of criticisms that this analogy is limited because the way a blanket works does not correspond well with the way greenhouse gases do. In addition, some researchers have also pointed out that the term *greenhouse effect*, as an analogy itself, is not exactly accurate considering the dissimilarities between a greenhouse and an atmosphere enriched with carbon dioxide [e.g., *Glantz*, 1991].

Therefore, when providing analogies for the public, scientists should first scrutinize them carefully and make sure they are structurally faithful to the topics in question. Besides, no one analogy is a perfect representation for the target, unfamiliar concept. Multiple analogies should be employed to help understand the target concepts and scientists should also consider pointing out the weaknesses of these analogies to the public. Furthermore, when teaching global climate change with analogies, scientists and educators should make sure that explanations of how the analogs work do not hamper understanding of global climate change. *Brousseau* [1997] argued that in the didactical relationship, target knowledge may disappear if a teacher oversimplifies questions to match students' prior knowledge. Students may become too dependent on the analogs and lack in-depth thinking about the target information. After all, analogies can be an effective tool for reasoning, but they are not the entire story. Therefore, analogies may be better learning aids for beginners, but as individuals gain deeper understanding of global climate change, more emphasis should be directed from the analogs to the complex science itself.

### 4.3.2. Evidence-Based Argumentation

Argumentation has been recognized as an essential aspect of scientific thinking and reasoning [*Duschl and Osborne*, 2002; *Erduran and Jimenez-Aleixandre*, 2008]. *Kuhn* [1993] stated that the forms of thinking and reasoning "can be rigorously defined within the framework provided by the structure of argument" (p. 333). Argumentation represents the social negotiation of claims and evidence [*Kuhn*, 1991], and it constitutes the practical purpose of reasoning [*Mercier and Sperber*, 2011]; rigorous reasoning helps individuals to devise and evaluate arguments intended to persuade others, which will then contribute to the effectiveness and reliability of communication. Thus, argumentation skills are critical in knowledge-building processes in which explanations are developed to make sense of data and then presented to a community of peers for critique, debate, and revision [*Driver et al.*, 2000].

Argumentation should be a pivotal element in climate change education. The public is constantly subject to a multitude of debates regarding global climate change [*King and Kitchener*, 2004]. Faced with all the perspectives and discussions surrounding them, individuals should be able to reason critically about the information they receive as well as generating solid arguments to support or refute claims. Moreover, as research has shown, teaching students to reason, argue, and think critically will enhance their conceptual learning [*Osborne*, 2010]. Enhancing argumentation skills thereby enhances climate literacy.

To promote argumentation in climate change education, scientists and educators should first provide open discussion opportunities about climate issues. *van de Kerkhof* [2006] proposed a dialogue approach to enhance the learning of climate science and suggested using "backward-analysis" (i.e., backcasting) to stimulate individuals to reflect on policy-making issues regarding global climate change. During backcasting, individuals are asked to consider and visualize a future image as a starting point for discussion and subsequently work backward to the present situation, collaboratively exploring which interventions are needed to realize this future. Because climate issues carry uncertainties that make it hard for individuals to predict or suggest actions for the future, the backcasting analysis breaks down the steps needed to achieve the goal and thus makes it less demanding to discuss and generate action plans. More importantly, it encourages individuals to actively engage in discussions where they can form and support their arguments in a social environment. Although this approach was originally designed for policy makers, it can be easily modified to apply in classrooms and the broader public.

Open discussion opportunities alone are not sufficient to enhance the quality of argumentation. Students have difficulties in learning how to engage in productive scientific argumentation to propose and justify an explanation [*Acar et al.*, 2010; *Sampson and Clark*, 2008]. To accurately argue about climates issues, students need to specifically learn about the types of claims that scientists make, how scientists advance them, what kinds of evidence are needed to warrant one idea over another, and how that evidence can be gathered and interpreted in terms of community standards [*Kelly and Chen*, 1999; *Sandoval and Reiser*, 2004]. Because learning these skills is challenging to not only children but also adults, explicit instruction on how to construct advanced argumentation is especially necessary [*Osborne et al.*, 2004].

Many curricula have adopted *Toulmin*'s argumentation pattern [1958] as a model and foundation for teaching argumentation on socio-scientific issues like global climate change. According to *Toulmin*, the statements that make up an argument have different functions, including claims, data, warrants, backings, qualifiers, and rebuttals. The strength of an argument is based on the presence or absence of specific combinations of these structural components [*Sampson and Clark*, 2008]. To improve the quality of argumentation, educators should first teach students how to generate each of these components and incorporate them into their arguments [*Bell and Linn*, 2000]. Because some of the components can be confusing, educators should use specific argumentation examples to teach students how to make a solid argument during the initial teaching and learning.

At the same time, activities such as writing refutable essays and participating in classroom debates can be helpful hands-on experience for students to practice argumentation after teachers' instruction. In these activities, students are encouraged to adopt different means to articulate and deliberate on the different viewpoints and ideas that exist in the group. They may also write down the evidence they would use to support their claims as well as how they would construct counterarguments. Through this process, they will gradually develop strategies to justify their theories, using the argumentative framework of alternative theories, counterarguments, and rebuttals, and eventually be able to make profound arguments about global climate change.

### 4.3.3. Epistemological Reasoning

For arguments to be considered persuasive and convincing, they must be consistent with the epistemological criteria used by the larger scientific community for "what counts" as valid and warrant scientific knowledge. Epistemological representations are an important explanatory mechanism that captures individuals' understanding of the nature of science. Such understanding involves, but is not limited to, views about what natural phenomena can or cannot be explained, what may count as a valid explanation, what can prove or disprove an explanation, and so on [*Driver et al.*, 1996]. Epistemological representations are usually embedded in scientific reasoning. This reasoning process is epistemological reasoning. Epistemological reasoning can be understood and evaluated from aspects such as individuals' approaches to exploration, depth of information processing, ability to deal with competing knowledge claims, and responses to anomalous data [*Tytler and Peterson*, 2003]. In other words, epistemological reasoning entails how individuals' understanding of the nature of science influences their scientific reasoning processes and outcomes [*Driver et al.*, 2000].

Limited epistemological reasoning may lead people to appreciate science in a limited way and thus be unlikely to see its relevance to their own lives. IPCC has stressed that "the evaluation of uncertainty and the necessary precaution is plagued with complex pitfalls. These include the global scale, long time lags between forcing and response, the impossibility to test experimentally before the facts arise, and the low frequency variability with the periods involved being longer than the length of most records" [*IPCCWG3*, 2001, p. 656]. It is thus suggested that climate change communications emphasize that "we cannot dismiss the possibility but we do not really know" when it comes to complex and ambiguous scientific evidence [*Charlesworth and Okereke*, 2009]. The development of appropriate epistemological reasoning promotes the appreciation of the complexity and uncertainty of climate science. With an accurate epistemology, individuals will avoid limiting their thinking to personal beliefs and be able to generate scientifically sound arguments. Besides, an accurate epistemology can help individuals to more objectively understand and evaluate scientists' expertise on climate science, which will in turn enhance the quality of their argumentation.

The public's epistemological representations toward controversial social issues have been discussed in *Kuhn*'s [1991] work, in which she identified three main epistemological theories in reasoning: *absolutist*, *multiplist*, and *evaluative theory*. It was suggested that absolutists believe that experts' knowledge is certain and absolute, and they either *do* or *can* know with certainty about the causes of the phenomenon in question. In contrast, multiplists and evaluative theorists deny the possibility of expert certainty. Multiplists recognize that knowledge does not only consist of facts and accept the coexistence of multiple viewpoints that can be equally correct. Similar to multiplists, evaluative theorists also acknowledge that viewpoints can be compared with one another and evaluated with respect to their relative adequacy or merit. They consider that although absolute certainty is impossible, experts may come closer to achieving such certainty than non-experts with close observation, examination, and analysis.

*Kuhn*'s [1991] categorizations capture the general epistemological representations individuals may hold. In climate change education, many people tend to explain climate science from an absolutist perspective, whereas full comprehension of the complexity and uncertainties requires multiplist perspectives and even evaluative theories. *Driver and colleagues* [1996] proposed a framework that taps more specifically into how epistemological representations can affect individuals' scientific reasoning. Based on the way individuals evaluate phenomena, scientific evidence, and theories, they proposed three qualitatively distinct representations, including phenomenon-based, relation-based, and model-based. Although those who hold a phenomenon-based epistemology do not distinguish between *description* and *explanation* of phenomena, a relation-based epistemology features explanations that take the form of relations between observable features, which can be either a chain of cause-and-effect relationships or linear causal reasoning. People who have a model-based epistemology, in contrast, express explanations in terms of a theoretical system, and view scientific inquiry as the evaluation of conjectured models in the light of evidence. They recognize that empirical evidence can never "prove" the truth of a conjectured model, although it can eliminate competing conjectures.

To further illustrate *Driver and colleagues'* [1996] framework, we will take responses to the question *"How would you explain the scientific processes of global climate change"* as an example. People who hold a phenomenon-based epistemology may simply say that "the Earth's temperature has been increasing in the past decades" without any further explanations, whereas those with a relation-based epistemology may explain the scientific processes of global climate change by saying "the Earth's temperature has increased dramatically since the industrial revolution because humans have released more greenhouse gases". In contrast, model-based epistemology reasoners are likely to provide in-depth explanations about how the water and carbon cycles work on the Earth, and they may also specifically include details about the greenhouse effect.

The categorizations *Kuhn* [1991] and *Driver and colleagues* [1996] proposed shed light on our exploration of why increased knowledge may not necessarily lead to enhanced climate literacy: when individuals consider scientific inquiry as direct observations of phenomenon and explain it in absolutist statements, no matter how much information is accessible, they will not develop advanced climate literacy. To improve the quality of epistemological reasoning is thus an urgent task, and yet not many studies have explored specific educational approaches that can be taken to serve this goal [*Khishfe and Abd-El-Khalick*, 2002]. Within the current literature, a broadly held view is that deep engagement in knowledge-building activities is the most likely path to promoting epistemological reasoning. *Sandoval* [2005] suggested engagement in authentic scientific inquiry as a means of developing epistemological understanding in climate science, and we have found it an effective approach in our work with in-service teachers. In the NASA-funded project we are currently working on, teachers participate in professional development workshops that aim to enhance their climate literacy and classroom teaching of global climate change [*Roehrig et al.*, 2012]. During the workshops, environmental scientists and climatologists are invited to explain scientific data collection processes and engage teachers in hands-on inquiry to collect and analyze climate data. As teachers work collaboratively to explore scientific evidence of local climate change, they come to a more in-depth understanding of the nature of the scientific processes behind climate science. More importantly, they grow to better recognize the uncertainty involved in such scientific processes and are more open to alternative explanations.

In climate change education for the broader public, scientists and educators should acquire firsthand information regarding the nature of people's epistemology during reasoning about climate issues. The evaluation can be in the form of public survey and interviews, and potential dimensions to be investigated include individuals' perspectives toward scientists' expertise, their view of the certainty, simplicity and source of knowledge, as well as the way they justify what they know and how they evaluate their own knowledge and that of others [*Hofer and Pintrich*, 1997]. Based on such information, scientists and educators will then be able to identify common concerns and doubts about the scientific practices, and thus frame scientific findings in a way to specifically avoid common misconceptions about scientific work on climate issues. Furthermore, scientists should also consider providing more details about the scientific inquiry processes that lead to findings and alternative interpretations. Detailed explanations in this regard can make it more likely for individuals to incorporate the new information into their existing knowledge and re-evaluate their understanding of climate science. This way, individuals' epistemological reasoning will gradually enhance and then facilitate an improvement in their argumentation skills and eventually climate literacy [*Weinstock*, 2005].

## 4.4. CONCLUSIONS

Global climate change is a problem that calls not only for social awareness, but also individual actions [*Feierabend and Eilks*, 2010]. Enhancing the public's climate literacy is an interdisciplinary issue that requires scientists and educators to work together to synthesize and share the scientific knowledge of an immensely complex system encompassing the climate, the economy, and society [*Sterman*, 2011]. With the growing body of scientific findings for global climate change, there is a growing gap between the scientific and public understanding about this issue [*Gallup*, 2010]. Because many factors may exacerbate this gap, such as the complexity of climate science and personal ideological orientations, there is an immediate need for the discussion of cognitive processes that may promote the understanding of complex climate science. This chapter introduced scientific reasoning into the endeavor of enhancing the public's climate literacy.

As a crucial cognitive mechanism, scientific reasoning is essential in not only scientific inquiry but also in our thinking and learning in everyday lives. In the context of climate change education, developing appropriate scientific reasoning skills will facilitate conceptual understanding of climate science, encourage critical thinking of public media messages, and enhance the quality of information communication about global climate change issues in the general public. Therefore, given its importance, we have emphasized three aspects of scientific reasoning and their importance in enhancing climate literacy. First, we suggested that scientists develop and employ more effective analogies as a reasoning tool to help the public coordinate their prior knowledge and the growing scientific evidence of global climate change. Because good analogies make it easier for people to relate the complex science to things they are familiar with, reasoning with analogies should be paid more attention to in climate change education. Second, we focused on the importance of encouraging more evidence-based argumentation in the discussions of climate issues. Scientists and educators should provide opportunities to enhance argumentation skills in the classrooms and strengthen its relationship with climate change education. Last, we discussed how epistemological representations may influence reasoning processes and their role in helping people to fully understand climate science. Developing an accurate epistemological representation will help individuals to avoid absolutist perspectives when faced with the complexity and uncertainty of climate science.

One of the most critical educational objectives is for students to learn about how socio-scientific issues are handled and evaluated within society so as to be able to act as responsible citizens in the future [e.g., *Höttecke et al.*, 2010. Incorporating scientific reasoning into climate change education will help fulfill this goal. It is essential that the general public come to appreciate the relevance of scientific reasoning and its impact on climate literacy. And to facilitate scientific reasoning and climate literacy, it is crucial that scientists and educators collaborate closely to provide more scientific information on global climate change that is compatible with the public's cognitive processes and encourage more discussions of global climate change in society as well as situating school science in such discussions.

We have by no means provided in this chapter an exhaustive account of what needs to be done to incorporate scientific reasoning into climate change education. However, by sampling from the three broad areas of scientific reasoning, we convey an idea of the necessarily wide scope required in detailing the relationship between scientific reasoning and climate change education.

## REFERENCES

Acar, O., L. Turkmen, and A. Roychoudhury (2010), Student difficulties in socio-scientific argumentation and decision-making research findings: Crossing the borders of two research lines, *Int. J. Sci. Ed.*, 32 (9), 1191–1206.

Anderson, J. R. (1990), *Cognitive Psychology and its Implications*, (3rd ed.), Freeman, New York.

Bauer, M., A. Allum, and S. Miller (2007), What can we learn from 25 years of PUS survey research? Liberating and expanding the agenda, *Public Understanding of Science*, 16, 79–95.

Bell, P., and M. C. Linn (2000), Scientific arguments as learning artifacts: Designing for learning from the web with KIE, *Int. J. Sci. Ed.*, 22 (8), 797–818.

Bostrom, A., and D. Lashof (2007), Weather or climate change? In S. Moser and L. Dilling (Eds.), *Creating a Climate for Change: Communicating Climate Change and Facilitating Social Change* (p. 31–43), Cambridge University Press, Cambridge, United Kingdom.

Brousseau, G. (1997), *Theory of Didactical Situations in Mathematics*, Kluwer Academic Publishers, Dordrecht, the Netherlands.

Chan, J., S. Paletz, and C. D. Schunn (2012), Analogy as a strategy for supporting complex problem solving under uncertainty, *Memory and Cognition*, 40, 1352–1365.

Charlesworth, M., and C. Okereke (2009), Policy responses to rapid climate change: An epistemological critique of dominant approaches. *Global Environmental Change*, 20 (1), 121–129.

Cialdini, R. (2009), *Influence: Science and Practice* (5th ed), Pearson, Boston.

Dolan, E., and J. Grady (2010), Recognizing students' scientific reasoning: A tool for categorizing complexity of reasoning during teaching by inquiry, *J. Sci. Teacher Ed.*, 21, 31–55.

Driver, R., J. Leach, R. Millar, and P. Scott (1996), *Young People's Images of Science*, Open University Press, Milton Keynes, United Kingdom.

Driver, R., P. Newton, and J. Osborne (2000), Establishing the norms of scientific argumentation in classrooms. *Sci. Ed.*, 84(3), 287–312.

Dunbar, K. (1997), How scientists think in the real world: Implications for science education. *J. Appl. Dev. Psych.*, 21 (1), 49–58.

Dunbar, K. (2001), The analogical paradox: Why analogy is so easy in naturalistic settings, yet so difficult in the psychology laboratory, in D. Gentner, K. J. Holyoak, and B. Kokinov (Eds.), *Analogy: Perspectives from Cognitive Science* (pp. 313–334), MIT Press, Cambridge, MA.

Duschl, R., and J. Osborne (2002), Supporting and promoting argumentation discourse in science education. *Stud. Sci. Ed.*, 38, 39–72.

Einstein, A. (1954), *Ideas and Opinions*, Crown, New York.

Ennis, R. H. (1987), A taxonomy of critical thinking dispositions and abilities, in J. B. Baron and R. J. Sternberg (Eds.), *Teaching Thinking Skills: Theory and Practice* (p. 9–26), W. H. Freeman, New York.

Erduran, S., and M. Jimenez-Aleixandre (2008), *Argumentation in Science Education: Perspectives from Classroom-based Research*, Springer, Dordrecht, the Netherlands.

Feierabend, T. and I. Eilks (2010), Raising students' perception of the relevance of science teaching and promoting communication and evaluation capabilities using authentic and controversial socio-scientific issues in the framework of climate change, *Sci. Ed. Int.*, 21 (3), 176–196.

Gallup (2010), Americans' global warming concerns continue to drop. Retrieved November 1, 2013, from www.gallup.com/poll/126560/americans-global-warming-concerns-continue-drop.aspx.

Gentner, D., and A. B. Markman (1997), Structure mapping in analogy and similarity. *Am. Psychs.*, 52, 45–56.

Giere, R., J. Bickle, and R. Mauldin (2006), *Understanding Scientific Reasoning*, Wadsworth-Cengage Learning, Belmont, CA.

Glantz, M. (1991), Use of analogies in forecasting ecological and societal responses to global warming, *Environment*, 33(5), 11–33.

Glynn, S. M., and T. Takahashi (1998), Learning from analogy-enhanced science text, *J. Res. Sci. Teach.*, 35, 1129–1149.

Halford, G. S., and P. W. Sheehan (1991), Human response to environmental changes, *Int. J. Psych.*, 26(5), 599–611.

Harrison, A. G. (2008), Teaching with analogies: Friends or foes? In A. G. Harrison and R. K. Coll (Eds.), *Using Analogies in Middle and Secondary Science Classrooms* (pp. 6–21), Cornwall Press, Thousand Oaks, CA.

Hofer, B. K., and P. R. Pintrich (1997), The development of epistemological theories: Beliefs about knowledge and knowing and their relation to learning. *Rev. Educ. Res.*, 88–140.

Höttecke, D., C. Hössle, I. Eilks, J. Menthe, M. Mrochen, H. Oelgeklaus, et al. (2010), Judgment and decision-making about socio-scientific issues: A fundament for a cross-faculty approach towards learning about climate change. In I. Eilks and B. Ralle (Eds.), *Contemporary Science Education* (pp. 179–192), Shaker, Aachen.

Houghton, J. T., L. G. Meira Filho, B. A. Callander, N. Harris, A. Kattenberg, and K. Maskell (1996), *Climate Change 1995: The Science of Climate Change*, Cambridge University Press, New York.

Inhelder, B., and J. Piaget (1958), *The growth of logical thinking from childhood to adolescence*, Basic Books, New York.

Intergovernmental Panel on Climate Change (IPCC) (2007), Summary for policymakers, in S. Solomon D. Qin, M. Manning, Z. Chen, M. Marquis, K. B. Averyt, M. Tignor and H. L. Miller (Eds.), *Climate Change 2007: The Physical Science Basis. Contribution of Working Group I to the Fourth Assessment Report of the Intergovernmental Panel on Climate Change*, Cambridge University Press, Cambridge, United Kingdom and New York.

IPCCWG3 (2001), *Climate Change 2001: Mitigation*, Cambridge University Press, Cambridge.

Kearney, A. (1994), Understanding global change: A cognitive perspective on communicating through stories, *Clim. Change*, 27, 419–441.

Kellstedt, P., S. Zahran, and A. Vedlitz (2008), Personal efficacy, the information environment, and attitudes toward global warming and climate change in the United States, *Risk Anal.*, 28(1), 113–126.

Kelly, G. J., and C. Chen (1999), The sound of music: Constructing science as a sociocultural practice through oral and written discourse, *J. Res. Sci. Teach.*, 36 (8), 883–915.

Khishfe, R., and F. Abd-El-Khalick (2002), Influence of explicit and reflective versus implicit inquiry-oriented instruction on sixth graders' views of nature of science, *J. Res. Sci. Teach.*, 39(7), 551–578.

King, P. M., and K. S. Kitchener (2004), Reflective judgment: Theory and research on the development of epistemic assumptions through adulthood. *Educ. Psych.*, 39, 5–18.

Kuhn, D. (1991), *The Skills of Argument*, Cambridge University Press, Cambridge, United Kingdom.

Kuhn, D. (1993), Science as argument: Implications for teaching and learning scientific thinking, *Sci. Ed.*, 77 (3), 319–337.

Lawson, A. (2005), What is the role of induction and deduction in reasoning and scientific inquiry, *J. Res. Sci. Teach.*, 42 (6), 716–740.

Leiserowitz, A., E. Maibach, C. Roser-Renouf, and N. Smith (2011), *Climate change in the American Mind: American's global warming beliefs and attitudes in May 2011*. Yale University and George Mason University, Yale Project on Climate Change Communication, New Haven, CT.

Lemke, J. L. (1990), *Talking Science: Language, Learning, and Values*, Ablex Publishing Corp., Norwood, NJ.

Lucarini, V. (2004), Towards a definition of climate science, Atmos. *Oceanic Phys.*, 18 (5), 413–422.

Maibach, E., C. Roser-Renouf, and A. Leiserowitz (2008), Communication and marketing as climate change intervention assets: A public health perspective, *Am. J. Prev. Med.*, 35, 488–500.

Means, M. L. and J. F. Voss (1996), Who reasons well? Two studies of informal reasoning among students of different grade, ability, and knowledge levels, *Cognition Instruct.*, 14, 139–178.

Mercier, H., and D. Sperber (2011), Why do humans reason? Arguments for an argumentative theory, *Behav. Brain Sci.*, 34, 57–111.

Moxnes, E., and A. Saysel (2008), Misperceptions of global climate change: Information policies, *Clim. Change*, 93, 15–37.

National Oceanic and Atmospheric Administration (NOAA) (2009), *Climate Literacy: The Essential Principles and Fundamental Concepts* (2nd ed.), retrieved November 1, 2013, from http://oceanservice.noaa.gov/education/literacy/climate_literacy.pdf.

National Research Council (2012), *A framework for K–12 science education: Practices, crosscutting concepts, and core ideas. Committee on a conceptual framework for new K–12 science education standards*. Board on Science Education, Division of Behavioral and Social Sciences and Education, The National Academics Press, Washington, DC.

Orgill, M., and G. Bodner (2005), The role of analogies in chemistry teaching, in N. Pienta, M. Cooper, and T.Greenbowe (Eds.) *Chemists' Guide to Effective Teaching* (pp. 90–105). Prentice-Hall, Upper Saddle River, NJ.

Osborne, J. (2010), Arguing to learn in science: The role of collaborative, critical discourse. *Science*, 328 (23), 463–466.

Osborne, J., S. Erduran, and S. Simon (2004), Enhancing the quality of argumentation in science classrooms, *J. Res. Sci. Teach.*, 41 (10), 994–1020.

Partnership for 21st Century Skills (2006), *A State Leader's Action Guide to 21st Century Skills: A New Vision for Education*, Partnership for 21st Century Skills, Tucson, AZ.

Roehrig, G., K. Campbell, D. Dalbotten, and K. Varma (2012), CYCLES: A Culturally-relevant Approach to Climate Change Education in Native Communities, *J. Curriculum Instruct.*, 6, 73–89.

Sadler, T. (2009), Socioscientific issues in science education: Labels, reasoning, and transfer, *Cult. Stud. Sci. Ed.*, 4, 697–703.

Sadler, T., S. Barab, and B. Scott (2007), What do students gain by engaging in socioscientific inquiry, *Res. Sci. Ed.*, 37, 371–391.

Sadler, T., and L. Donnelly (2006), Socioscientific argumentation: The effects of content knowledge and morality, *Int. J. Sci. Ed.*, 28 (12), 1463–1488.

Sampson, V., and D. Clark (2008), Assessment of the ways students generate arguments in science education: Current perspectives and recommendations for future directions, *Sci. Ed.*, 92(3), 447–472.

Sandoval, W. A. (2005), Understanding students' practical epistemologies and their influence on learning through inquiry, *Sci. Ed.*, 89 (4), 634–656.

Sandoval, W. A., and B. J. Reiser (2004), Explanation driven inquiry: Integrating conceptual and epistemic scaffolds for scientific inquiry, *Sci. Ed.*, 88 (3), 345–372.

Schauble, L. (1996), The development of scientific reasoning in knowledge-rich contexts, *Dev. Psych.*, 32(1), 102–119.

Schneider, S. H. (2007), http://cesp.stanford.edu/news/cesp_codirector_and_climate_scientist_stephen_schneider_testifies_before_congress_on_the_subject_of_climate_change_risks_and_control_strategies_20070327/index.html.

She, H., and Y. Liao (2010), Bridging scientific reasoning and conceptual change through adaptive web-based learning, *J. Res. Sci. Teach.*, 47 (1), 91–119.

Sterman, J. (2011), Communicating climate change risks in a skeptical world, *Clim. Change*, 106, 611–626.

Taylor, F. W. (1991), The greenhouse effect and climate change, *Rep. Prog. Phys.*, 54(6), 881–918.

Toulmin, S. (1958), *The Uses of Argument*, Cambridge University Press, Cambridge, United Kingdom.

Tweney, R. D. (1991), Informal reasoning in science, in J. F. Voss, D. N. Perkins, and J. W. Segal (Eds.), *Informal Reasoning and Education* (p. 3–16), Erlbaum, Hillsdale, NJ.

Tytler, R., and S. Peterson (2003), Tracing young children's scientific reasoning, *Res. Sci. Ed.*, 33, 433–465.

van de Kerkhof, M. (2006), A dialogue approach to enhance learning for sustainability-A Dutch experiment with two participatory methods in the field of climate change, *Integr. Assess. J.*, 6(4), 7–34.

Wason, P. C., and P. N. Johnson-Laird (1972), *Psychology of reasoning: Structure and Content*, Batsford, London, England.

Weinstock, M. P. (2005), Cognitive bases for effective participation in democratic institutions: Argument skill and juror reasoning. *Theor. Res. Soc. Educ.*, 33, 73–103.

# 5
## Evaluation and Assessment of Civic Understanding of Planet Earth

### Julie C. Libarkin*

This book argues for civic understanding of the Anthropocene—understanding that encompasses the physical setting of planet Earth, the interactions between humans and the Earth that are dramatically shaping our planet, and the challenges that humans face in trying to understand and respond to rapid environmental change. The nature of cognitive processes, the complex phenomena that occur within the human mind, dictates that responding to our changing environment cannot and will not be a simple process. Even more difficult, human cognition must be considered within the framework of society. Fundamentally, civic understanding of natural phenomena is important because of the role that individual knowledge and affect play in decision making and ultimately behavior toward the environment. Many scholars in recent decades [*Stapp*, 1970; *Hungerford et al.*, 1980; *Volk*, 1984; *Grace*, 2009; *Nicolaou et al.*, 2009] have argued that effective environmental education should result in a citizenry engaged in solving the world's environmental problems. This engagement can take many forms, from simply supporting environmental initiatives to being involved in stewardship activities. Regardless of the extent of involvement, an environmentally literate populace must be able to function as individuals interacting with the world and as members of societies making decisions about complex, global environmental issues.

This chapter will consider the importance of evaluation and assessment for any effort to educate, with specific focus on environmental, geoscience, Earth systems, and related educational efforts. The discussion will introduce basic concepts related to assessment, provide an overview of norms within different assessment communities, position assessment within a broader framework of pedagogical practice, and offer both practical and theoretical recommendations for anyone interested in characterizing civic understanding or assessing progress in response to education and outreach.

### 5.1. DEFINING EVALUATION AND ASSESSMENT

The terms *evaluation* and *assessment* can evoke heated debate among some members of the testing (or psychometric) community. For some scholars, the two words are synonymous and easily interchangeable, whereas others use each word carefully and in context-specific ways. For our purposes, this chapter will adopt *assessment* as an overarching umbrella term. In this framework, evaluation is the assessment of student learning that occurs within an instructional context. This can include formative (evaluation during learning) or summative (evaluation after learning) practices for measuring student learning. It might also include pre-assessments initiated to gauge student readiness to learn or to structure pedagogical practices to most effectively meet student needs. Research assessment, on the other hand, is undertaken to measure student learning in ways that contribute to both the local instructional context and the larger community engaged with scholarly discourse about student learning. Assessment of student learning for research purposes must by definition be held to higher standards than assessment for instruction. Ideally, assessment undertaken for instruction can double as research assessment, although the time-intensive nature of developing high-quality assessments, as would be needed for high-quality research, is something that is simply not feasible to expect from many instructors. Thankfully, the research community has come a long way in developing assessments that can be adapted for classroom purposes.

---

*Geocognition Research Lab, Department of Geological Sciences, Michigan State University, East Lansing, Michigan

*Future Earth—Advancing Civic Understanding of the Anthropocene, Geophysical Monograph 203*, First Edition.
Edited by Diana Dalbotten, Gillian Roehrig, and Patrick Hamilton.
© 2014 American Geophysical Union. Published 2014 by John Wiley & Sons, Inc.

## 5.2. THE IMPORTANCE OF EVALUATION AND ASSESSMENT FOR CIVIC UNDERSTANDING INITIATIVES

Science is fundamentally about evidence. This evidence provides support or refutation for hypotheses, or opens up new worlds for scientists to explore and question. Evidence is never used as "proof"; that is, scientists recognize that nothing can ever be absolutely known or understood. We can never observe everything, we will never be able to explore everywhere, and the possibility of a new discovery shaking our basic understanding of how the world works is always present. Science grows through both small discoveries and world-changing ones, and without evidence, our understanding of the world would be limited to personal observation and experience. At the same time, scientists are comfortable with the idea that science is imperfect, evolving, and changing in response to new evidence. Although this reality may make it difficult to communicate science adequately to the public, scientists understand and recognize the role of uncertainty in science [e.g., *Rodriguez et al.*, 2007].

Assessment of student learning is also fundamentally about evidence. Although we can never perfectly measure what is occurring in someone's mind, within a classroom sample of students, or across a population of museum visitors, we can collect evidence for or against the efficacy of education. This evidence will never constitute proof that instruction is effective, but rather offers opportunities for us to consider what it means to engage in best practices for learning. Certainly, assessment data can be used to identify situations in which learning seems to be occurring, and allows us to draw generalizations about the types of instruction that may be most useful for specific types of learners, with specific types of needs, in specific types of settings. Without evidence, educators are left to draw anecdotal conclusions about what works in instruction. Without a foundation of evidence, instruction cannot effectively transform in step with scientific discoveries.

Building a community through which educators can disseminate best practices for teaching about human impacts on the planet has the potential to transform how and what we teach in settings ranging from elementary schools to news broadcasts. Dissemination is vital to any scholarly endeavor, and in many ways, work is not completed until it is distributed to the broader community of practice. Assessment plays a vital role in helping us to determine what to disseminate; only those ideas, materials, and practices that are known to be effective should be distributed. From this perspective, we then must consider what constitutes evidence, how that evidence can be collected practically, and what it means for something to rise to the level of best practice. Certainly, all educators should ask themselves: "How can I know my practice is effective?" Data collected to understand efficacy can and should inform practice itself, effectively allowing educators to close the assessment loop and use assessment data to modify instruction and materials for improved student learning.

## 5.3. COGNITIVE, AFFECTIVE, AND BEHAVIORAL CONSIDERATIONS

*Wiggins and McTighe* [2001] lay out a framework for instructional design that is both beautiful in its simplicity and difficult to follow in practice. Backward Design is a process that encourages instructors to take a step back and think deeply before developing pedagogy and curriculum. The applicability of Backward Design to any setting, from elite high school AP courses to kindergarten classrooms to modern museums, rests on a common foundation that centers on asking what people should reasonably expect to get out of spending time engaged within an educational setting. Although the specific questions asked by a faculty teaching an undergraduate environmental science course might be different from the questions asked by developers of a television children's special, the fundamental structure is the same: "What should students be able to do, think, believe, understand, reason about, identify with, value, discuss, or argue after engaging in instruction?" This question requires instructors and designers to be explicit in defining anticipated learning outcomes, something that is surprisingly foreign to many educators. The importance of identifying learning outcomes cannot be overstressed. Although other chapters in this book (most notably in Chapter 2) lay out important criteria for environmental literacy, a brief review of anticipated outcomes for civic understanding follows, setting the basis for assessment practice discussed later in this chapter. No single educational initiative will address more than a small handful of these learning outcomes, let alone all of them. The list serves as a valuable starting point for considering the range and breadth of assessment practices that will be required as we move toward a better understanding of the impact and value of environmental and similar education on students and participants.

### 5.3.1. Learning Outcomes

Learning outcomes can be categorized in any number of ways; the approach of considering cognitive, affective, and behavioral outcomes is most valuable here specifically because decision making is intertwined with both cognition and affect, and decision making about the environment is needed for the behavioral changes discussed elsewhere (Chapter 4). Thus, understanding cognition and affect is a prerequisite for developing instructional interventions that will result in effective decision making and hopefully behavioral change [*Schwarz*, 2000].

Several documents have articulated learning outcomes that align with the notion of civic understanding of the Anthropocene, most notably those documents emerging from communities focused on environmental science and related fields. For example, sets of Literacy Principles have been developed for Earth, Ocean, Atmosphere, and Climate sciences (see http://eo.ucar.edu/asl/ for links to all four documents). These principles may focus exclusively on cognitive outcomes (e.g., Earth Science) or may offer blends of cognitive and affective outcomes (e.g., Climate). *Ladue and Clark* [2012] have synthesized these four distinct documents into a set of 11 overarching principles, several of which are particularly relevant to environmental education. Similarly, frameworks for science learning in K–12 classrooms lay out overarching themes; for example, the US science standards articulate crosscutting themes that broadly apply to all science disciplines and certainly to learning about human impacts on Earth [Chapter 2; *National Research Council (NRC)*, 1996, 2012]. Professional societies focusing on education have also developed their own sets of desired learning outcomes (e.g., http://www.naaee.net/framework). In general, outcomes described in each of these documents align well with the categories of cognition, affect, and decision making considered here and should be carefully reviewed during the goals-development phase of Backward Design.

### 5.3.1.1. Cognition

In this context, cognition refers to the skills and knowledge that are required for someone to be considered literate in the science of the environment and able to access and use information. The majority of the learning outcomes articulated for K–12 students or the general public focus on cognitive outcomes. For example, the NRC [2012] identifies understanding of Systems and Systems Models as an important outcome for students, whereas *Ladue and Clark* [2012] note that change in Earth systems is a major theme that cuts across the general literacy principles. Skills also play an important role in the existing standards, with the ability to explore Earth's systems through observation, modeling, and reasoning, whether directly or indirectly through information seeking, visible across documents. Fundamentally, the specific content taught during environmental education is perhaps less important than the need for students to gain an awareness of how the Earth works and the role humans play on the planet; this includes a need for learners to be able to reason about human-Earth interactions (Chapter 4). Assessment of cognition often uses standardized tests of conceptual understanding or specific skill, such as the Earth Science Regents Exam used by New York State for high school graduates or the tests used by the National Assessment of Educational Progress. The current section briefly summarizes the kinds of learning that can be assessed, whereas specific examples of tests that could be used are discussed later in this chapter.

### 5.3.1.2. Affect

Affect refers to the noncognitive variables of emotion, such as attitude, feeling, and value. This domain considers the ethics of individual impacts on the environment, for example. Assessment of affective variables is more complicated than assessment of cognition. Although skills and knowledge have absolute answers related to whether or not someone understands knowledge or can engage in a specific skill, affective variables rarely have right-or-wrong answers. Rather, assessment must focus on categorizing populations, documenting change in affect in response to instruction. We can hope that instruction will, for example, improve valuation of the environment or encourage more positive attitudes about environmental laws, but how an individual chooses to feel about the world is not something that can be dictated. Rather, educators should seek to help learners recognize that opinions and emotions can, and in fact, should be varied, and that these affective responses should be based on a strong understanding of how the world works.

### 5.3.1.3. Decision Making

Consideration of decision making in environmental education aligns well with approaches used in other fields, particularly in science [see *Sadler*, 2004 for a review]. Generally, the decision-making process is of most interest rather than the decision itself, specifically because, like affect, decisions are highly individual and cannot not be considered to be correct or incorrect [*Aikenhead*, 1985; *de Jager and van der Loo*, 1990; *Grace*, 2009]. Rather than seeking specific responses, the quality of argumentation [*Driver et al.*, 2000; *Kuhn et al.*, 1997], the role that educational interventions play in changing how students support decisions about complex topics [e.g., *Grace*, 2009], and the nature of group decision making [e.g., *Aikenhead*, 1989] are vital assessment outcomes in environmental education. Scholars can also compare individual traits, such as demographics, attitudes, conceptual understanding, and morality, to decision making to tease out important differences in the impacts of educational environments on different kinds of learners [e.g., *King and Mayhew*, 2002; also see Chapter 4].

## 5.4. A BRIEF REVIEW OF ASSESSMENT INSTRUMENT TYPES

As shown here and in other chapters (see Chapter 2, this volume), a wide array of learning outcomes can be linked to environmental, Earth science, and related instruction. Similarly, assessments developed to measure whether or not learning outcomes have been met are quite broad in style and scope, simply because a one-size-fits-all assessment approach will not effectively measure learning in all contexts. Some assessment approaches may be more appropriate for instructional-versus-research settings, and each approach offers unique benefits and drawbacks. This section provides an overview of the range of assessment types that can be implemented, with the hope that readers will be able to identify the approach that is most appropriate for their needs. Educators should adapt existing tools where possible to meet the specific needs of their unique settings and target learners. Of course, the development of new tools aligned with new or unevaluated learning outcomes is always needed. Readers are encouraged to develop new tools in alignment with best practices in instrument development [e.g., *DeVellis*, 2011].

Assessment can be characterized as qualitative, quantitative, or mixed method. Very simply, qualitative approaches result in anything except numbers, quantitative approaches result in numbers only, and mixed-methods approaches blend the two. In reality, all numerical data have some qualitative underpinnings and all qualitative information can be quantified in some way. Qualitative approaches generally provide a much deeper understanding of individual student thinking while providing limited information about larger populations. Quantitative approaches generate data that allow comparisons across many people, although providing limited information about individuals. Neither approach is better than the other; rather, the approach should be carefully chosen to align with the needs of the project being assessed as well as the resources available for the assessment itself, such as whether a trained assessment specialist is working on the project.

### 5.4.1. Qualitative Approaches

Qualitative data can take many forms, from verbal data in written or oral forms to visual data such as drawings or video. *Patton* [2002] is a comprehensive text that provides suggestions for how to collect qualitative data, approaches for analysis, and advice on which data are most appropriate for specific types of research questions. A summary of different forms of qualitative data is offered here; although not comprehensive, this summary covers approaches most likely to be used in educational settings (Table 5.1).

#### 5.4.1.1. Short Response Questions and Essays

Written responses to survey questions are perhaps the most common form of qualitative data collected in educational settings. Open-ended question prompts can be used to probe learner understanding or affective responses to instruction. Questions requiring short responses are quite easy to collect and analyze, and respondents are likely to respond with a few brief sentences. Essay-style questions requiring more detailed responses are less likely to be completed by many individuals, simply because long responses require more time and effort on the part of the learner. Written responses provide a record of learner thoughts, in their own words, although researchers are generally unable to ask learners to clarify what they have written or provide additional information.

**Figure 5.1** Visual of indexing of common terms used in previous two paragraphs. Created using http://www.wordle.net/.

Table 5.1 Summary of Qualitative Assessments Applicable to Educational Settings

| Approach | Description | Benefits | Drawbacks |
|---|---|---|---|
| **Short Answer** | Brief written responses to questions. | Can collect and analyze many responses in short time. | Responses are generally too short for deep understanding. Cannot probe for additional information. |
| **Essays** | Long written responses to questions. | Responses may provide insight into complex student thinking. | Time intensive to analyze. Rubrics for scoring are difficult to create and validate. Cannot probe for additional information. |
| **Drawings** | Nonverbal responses, typically to oral or written prompts. | Can collect many responses in short time. | Time intensive to analyze, although automated approaches are in development. |
| **Interviews** | Verbal responses to oral prompts from a single individual. | Responses may provide insight into complex student thinking. Responses can be probed to elicit additional information | Time intensive to collect and analyze |
| **Focus Groups** | Verbal responses to oral prompts from several individuals. | May provide insight into complex student thinking. Can probe to elicit additional information. | Probing of individuals is difficult. May lose individual nuances. |
| **Observations** | Documentation of activities in learning environments. Can include activities of learners as well as instructors or facilitators. May include permanent video record. | Influence of assessor on data is reduced if observer is passive (rather than active, as in some types of ethnography). May provide insight into complex student thinking. Can probe to elicit additional information. | Observations must be interpreted by researcher, generally without input from research subjects. |

Analysis of written responses can take several forms. Simple analysis might use the indexing approach to count words of importance. For example, a question probing learner understanding of river systems might count words related to stream flow and aquatic organisms. More complex indexing can look for relationships between words; for example, whether concepts related to river health and biodiversity occur together. Both simple and relational indexing can be automated and have been for years [e.g., *Salton et al.*, 1975; Fig. 5.1]. Other mechanisms for visualizing analyses of qualitative data, such as concept maps [e.g., *Novak*, 1990], are also quite useful for representing qualitative data.

Detailed content analysis of themes present within written responses provides a richer look into learner perspectives. Thematic content analysis looks for patterns within responses from a sample of learners. These patterns offer a mosaic of learner ideas that can be compared to expected learning outcomes and thus a sense of whether or not instruction is effective. Although indexing can be automated, thematic content analysis is more complicated than simple binary coding and requires human coders [e.g., *Nehm et al.*, 2012]. Establishing reliability of thematic content analysis is important and is generally done through simultaneous coding of responses by two or more researchers. Interrater reliability, a measure of the extent to which analytical results will be reproducible by other researchers, can then be reported as percentage overlap or through calculation of metrics such as the intraclass correlation.

### 5.4.1.2. Drawings

Use of drawings to understand learner thinking is common in environmental science [e.g., *Alerby*, 2000; *Bowker*, 2007]. The way in which learners visually represent their ideas, feelings, or behaviors can provide deep insight into how they perceive and value the world around them. In fact, researchers generally recommend that verbal data be coupled with drawings as a way to provide additional insight into student understanding [*Cheng and Gilbert*, 2007]. Analysis of drawings is time consuming in much the same way that analysis of verbal data is time consuming. Generally, approaches used to analyze drawings mirror those used for verbal response data— drawings can be sorted into common categories and components of drawings can be analyzed and interpreted.

Identification of common features in drawings is perhaps the most common approach used in drawing analysis [e.g., *Alerby*, 2000]. Computer-based pattern analysis of drawings is also viable [e.g., *Brown et al.*, 1987], although computer analysis is rarely used in assessment of student learning.

### 5.4.1.3. Interviews and Focus Groups

Interviews and focus groups provide an opportunity to interact with learners with the specific goal of probing learner understanding or perspectives [e.g., *Wilson*, 1997]. These data can be quite useful for identifying learner knowledge or for gaining insight into affect. Although some of these interactions may be highly scripted, such as with formally structured interviews, many interviews and focus groups are semi-structured and offer opportunities for the interviewer to probe in response to learner comments. As a consequence, interview and focus group data can offer detailed insight not possible through surveying approaches common with written or drawn data. Interviews are generally conducted with one individual at a time [e.g., *Cutter-Mackenzie and Smith*, 2003], whereas focus groups bring together groups of people to engage in a discussion [e.g., *Connell et al.*, 1999]. Analysis of verbalized data follows approaches used for written responses, although transcripts of interviews and focus groups are generally much longer than written responses. As a result, oral data sets generally take much longer to analyze than written data. In general, assessment of oral data uses responses from a much smaller sample than assessment using written or drawn survey data.

### 5.4.1.4. Observations

Observations of learners may be the most useful qualitative approach for collecting information about learner attitudes and behavior. Observations provide actual, firsthand data about learner interactions with the environment, rather than asking learners to report on those interactions. Observation can be used during instruction to document how educators are interacting with learners, to detail actions of learners, to understand how learners are interacting with educational products, such as museum exhibits, or to document how learners behave after instruction occurs. The record of observation can be a videotape that is subsequently analyzed [e.g., *vom Lehn*, 2006], or a formalized record of actions and interactions [e.g., *Sawada et al.*, 2002]. The approach used to analyze observational data depends on the extent to which the assessment needs to be comparable to other studies or unique to the educational setting under study.

### 5.4.2. Quantitative Approaches

Quantitative data collected in educational settings to assess student learning might be more properly termed *semi-quantitative* (Table 5.2). That is, numerical data collected from learners are generally in the form of scores on a survey or test, scores that depend strongly on the specific question being asked. The numbers generated are not exact measures, but are rather suggestive of learner understanding, affect, or behavior. This discussion focuses on two types of survey questions used commonly in education: Likert-type and multiple-choice. Both Likert-type and multiple-choice tests are easy to administer and analyze, offering the ability then to collect data from a large sample and the ability to compare results across different populations and settings. The ability to analyze Likert and multiple-choice data statistically can be deceiving, however; the data collected by these instruments is only as valid as the instrument itself. Careful attention must be paid to instrument design and validation, a process that can be both time consuming and overwhelming [*DeVellis*, 2011].

### 5.4.2.1. Likert-Type Questions

Likert-type questions allow collection of ordinal data quickly and are commonly used to measure affective or behavioral outcomes. Generally, Likert questions require respondents to indicate the level to which they agree or disagree with a statement. For example, a student enrolled in a college-level course might be asked to indicate the degree to which they agree with the statement, "The instructor was easy to understand", at the end of the course. This agreement can be fixed (e.g., Strongly Agree, Agree, Disagree, Strongly Disagree) or can be represented as a sliding scale. For example, Likert-type instruments have been used for decades to measure environmental perspectives and worldviews [*Dunlap et al.*, 2000; Fig. 5.2].

Although responses to individual questions can be interesting, the totality of responses across several questions usually provides the most insight into learners. Often, responses are given numerical values, such as 4 for Strongly Agree and 1 for Strongly Disagree, and an overall score is calculated as the sum or average across

---

Humans are severely abusing the environment.

STRONGLY AGREE, AGREE, UNDECIDED, DISAGREE, STRONGLY DISAGREE

**Figure 5.2** Likert-type item from the New Ecological Paradigm Scale of *Dunlap et al.* [2000].

> Which of the following best describes the relationship between the greenhouse effect and global warming?
>
> A. The greenhouse effect and global warming are likely the same thing.
> B. Without the greenhouse effect, there would be almost no global warming.
> C. Without global warming, there would be almost no greenhouse effect.
> D. The greenhouse effect and global warming are likely unrelated.
> E. There is no definite proof that either the greenhouse effect or global warming exists.
> F. I do not know.

**Figure 5.3** Multiple-choice question from the Geoscience Concept Inventory. Archived at http://geoscienceconceptinventory.wikispaces.com/Relating+the+Greenhouse+Effect+and+Global+Warming+%28open%29.

**Table 5.2** Summary of Quantitative Assessments Applicable to Educational Settings

| Approach | Description | Benefits | Drawbacks |
| --- | --- | --- | --- |
| **Likert, fixed response** | For example, Strongly Agree to Strongly Disagree response options. | Quick to collect and analyze | Difficult to validate. |
| **Likert, sliding scale response** | For example, a line along which respondents are asked to place themselves. | Quick to collect and analyze | Difficult to validate. |
| **Multiple-choice, single response** | Fixed response answers with one "correct" answer | Quick to collect and analyze | Difficult to create and validate. |
| **Multiple-choice, multiple response** | Fixed response answers with one or more "correct" answers. | Data can be equivalent to data collected through short answers if response options represent authentic responses. | Difficult to create, analyze, and validate. |

many questions. Some Likert-type scales incorporate a "Neutral" or "Undecided" response in the middle of the scale (i.e., the scale above would be 5 points), whereas others force respondents to articulate a preference. Regardless, changes in scores over the course of instruction can then offer insight into changes in student affect or behavior that occur as a result of education.

*5.4.2.2. Multiple-Choice Questions*

Multiple-choice questions have a longstanding tradition in education, for use as classroom assessment tools and as measures of general educational achievement. The longstanding tradition of using multiple-choice questions for educational assessment has generated significant attention from test developers, resulting in well-established rules for best practice in multiple-choice item writing. Readers are encouraged to consider these standards [e.g., *Haladyna*, 2004] when evaluating the value of existing instruments or writing new instruments.

Multiple-choice questions most commonly come in the form of single-response option questions (Fig. 5.3), where one response option is the correct response. More rarely, assessment instruments will incorporate multiple-response option questions, requiring respondents to choose all responses that are correct. In most cases, a score on a multiple-choice instrument will be reported as a percentage correct or will be an ability score calculated from the percentage correct. Multiple-choice instruments are often used to measure learner cognition. For example, concept inventories are a type of multiple-choice instrument used in higher education to measure conceptual understanding; these instruments are discussed in more detail in the following section.

## 5.5. CURRENT ASSESSMENT PRACTICES IN GEOSCIENCE EDUCATION

This section briefly reviews the current status of assessment practice in environmental, geoscience, Earth systems science, and related disciplines. Although not exhaustive, readers are provided with guidance toward existing assessment instruments or assessment sites that may be useful in their own projects. These resources are considered here in relation to the learning outcomes of cognition and affect, specifically because these outcomes are much more commonly measured than behavior or decision making. Where possible, reference is made to the population (e.g., K–12, undergraduate, informal learners) for whom materials were designed, although many instruments can be adapted for different populations. For example, instruments developed for school or university settings are certainly appropriate for use in informal settings, particularly those instruments designed for younger students or non-scientists. Users should carefully consider the learning outcome being tested as well as the appropriateness of an instrument for a specific population before choosing an assessment.

### 5.5.1. Cognition

Assessments done on the national scale tend to focus on cognitive outcomes related to traditional school subjects. For example, the National Assessment of Educational Progress (NAEP; Table 5.3) focuses on reading, mathematics, science, civics, and similar topics. Although geography is a core area, the focus on core school subjects on the NAEP means that few questions consider student understanding of the complex role of human impacts on the Earth. At a similar national level, the American Association for the Advancement of Science (AAAS) has recently created a concept assessment site through which measurement of middle- and high school student learning can occur (see Table 5.3). Tied to national standards, these assessments offer questions that align more tightly with the types of environmental subjects that are of interest when considering how best to measure civic understanding of the Anthropocene. In addition to national exams, assessments of knowledge have been developed for use at the state level, usually for teacher accreditation or high school graduation requirements. The New York State Regents Exams (see Table 5.3) cover four core science areas, including Earth Science and the Living Environment. Regents exams have been administered for more than 100 years, with many of the exams administered in previous years available online. Regents exams are generated by groups of experienced assessment writers and teachers, resulting in high-quality questions that can serve as a strong basis for measurement of learner knowledge.

Institutions of higher education have much more flexibility in what they teach than K–12 settings, and by extension in how they assess student learning. The past half-century has seen the development of a set of assessment instruments that have been used to understand undergraduate student cognition, affect, and behavior as it relates to the environment. These instruments emerge from within the scientific community or from groups specifically trained in testing. As a result, the underlying rationale for these assessments as well as their validity and reliability are not always clear.

A suite of multiple-choice instruments known as concept inventories have emerged in the last two decades as scientists have become more interested in understanding exactly what students are learning in college, and occasionally high school, classrooms. Concept inventories are intended to specifically target student alternative conceptions about science; generally speaking, incorrect response options are generated from student ideas documented in the research literature or identified by concept inventory developers themselves. Theoretically, a student's choice of a specific response option would then provide insight into student thinking much more powerfully than standard tests and can illuminate barriers to student learning. In practice, concept inventories are developed with a varying degree of attention to psychometric principles of test design [*Libarkin and Jardeleza*, 2013]. Several concept inventories are particularly relevant to civic understanding of the Anthropocene, specifically those that focus on climate change processes, environmental science, or geoscience. These include the Geoscience Concept Inventory [*Libarkin and Anderson*, 2006], The Greenhouse Effect Concept Inventory [*Keller*, 2006], and a set of questions related to climate change under development [*Jarrett et al.*, 2012].

### 5.5.2. Affect

Two online resources, PEARweb and the FLAG Guide (see Table 5.3), offer links to a number of instruments that measure affect about science or the environment. The Program in Education, Afterschool, and Resiliency (PEAR) has created a clearinghouse for assessment instruments that have been used in a wide variety of formal and informal educational settings. Those instruments used in informal settings are of particular interest for affective responses because many informal settings (zoos, museums, aquariums, etc.) focus education on changing individual emotions, attitudes, or beliefs rather than knowledge acquisition. The Assessment Tools in Informal Science (ATIS) site (http://www.pearweb.org/atis) has been designed to help educators identify assessment instruments that align with program needs. The database can be searched based on specific criteria such as learner age or target outcome; the site also offers ratings and reviews by users of instruments. Some instruments are available directly from the site whereas others must be requested from authors; a cursory perusal of instruments suggests that about three-quarters of the listed instruments are available directly online.

The Field-tested Learning Assessment Guide (FLAG) was designed to serve as a clearinghouse for higher education assessment instruments. Instruments included in

**Table 5.3** Assessment Websites of Relevance to Civic Understanding of the Anthropocene

| Website | Description and website |
|---|---|
| NAEP | The National Assessment of Educational Progress (NAEP): http://nces.ed.gov/nationsreportcard/. National and longstanding study of student progress in many subjects. Includes geography and science. |
| AAAS | The American Association for the Advancement of Science (AAAS) assessment project: http://assessment.aaas.org/. Provides multiple-choice questions and online assessment site. Aimed at K–12 science instruction and aligned with national standards. |
| GCI wiki | Geoscience Concept Inventory (GCI): http://geoscienceconceptinventory.wikispaces.com/. Project to bring together assessment questions written across the geosciences, broadly construed. Aimed at upper level secondary and university students. |
| PEARweb | Program in Education, Afterschool, & Resiliency (PEAR) assessment sites: http://www.pearweb.org/tools/. Project to bring together many tools for assessment in schools, classrooms, and informal settings. |
| FLAG | The Field-Tested Learning Assessment (FLAG) Guide: http://www.flaguide.org/. Project brings together instruments used in higher education to assess learning across science, technology, engineering, and mathematics. |
| NY Regents | The New York State Regents Exams in Science: http://www.nysedregents.org/regents_sci.html. Exams required of all high school graduates in New York state. Past exams are posted and include relevant topics in Earth Science and the Living Environment. |
| NAAEE | The North American Association for Environmental Education (NAAEE): http://www.naaee.net/framework. A set of documents describing the NAAEE's perspective on best practices for assessing environmental literacy, including dimensions of environmental literacy that should be targeted and assessed in instruction. |

the database cover both cognitive and affective outcomes, similar to PEAR, although no Earth or environmental cognitive instruments are listed. The assessment instruments focusing on affective variables, such as learner attitudes toward science, would be applicable or easily modifiable for assessment of affect in the context of the Earth. Although the project that created FLAG is no longer funded, the site is still active and assessment instruments can be downloaded or requested from authors.

Finally, the North American Association for Environmental Education (NAAEE) has laid out guidelines for assessment of environmental literacy that should be considered carefully by anyone working toward comprehending how the public understands and responds to Earth–human interactions. This assessment guide is aligned with the desired learning outcomes, or "competencies, knowledge, and dispositions," which NAAEE has laid out as part of its executive position on environmental literacy and education. The assessment framework articulated by NAAEE acknowledges the complexity of environmental literacy, including the interplay that occurs between cognition, affect, behavior, and decision making. Although assessment at the scale and complexity suggested by the NAAEE is likely outside the reach of educators conducting assessment for internal or small-scale purposes, any project that can undertake more complex assessment should look to the NAAEE plan as a useful framework that has been both developed and adopted by the larger community.

## 5.6. RECOMMENDATIONS FOR ASSESSMENT BEST PRACTICES

The status of assessment practices and opportunities relevant to civic understanding of the Anthropocene is fairly robust, suggesting that we have ample opportunity to engage in collection of meaningful data that can inform how we engage the public in thinking about our planet. As we move forward to build mechanisms for assessing learning across educational settings, several recommendations for best practice should be considered. These recommendations are geared toward ensuring that data are most relevant to modifying formal and informal education to improve learning and toward inclusive practices that acknowledge the diversity of learners who need to have facility in making decisions about Earth.

1. *Backward Design: Setting Learning Outcomes First.* Clearly articulating the purpose behind education, whether formal classroom instruction or informal experiences, is not always easy. In fact, the development of curriculum often precedes purposeful and explicit generation of desired learning outcomes, with assessment development occurring at the end of the process. In the real world, that is, designing a measure of student learning often occurs only when a test is needed to assign a grade or data are needed to ensure a funding agency that a project was successful. In many cases, then, assessment is ultimately poorly aligned with project needs, essentially because the anticipated impact on individual learners was never clearly identified. Spending time clarifying what

learners will gain from an experience is vital, as is considering cognitive, attitudinal, and behavioral outcomes explicitly. Often, the careful outlining of what is expected from educational experiences leads to more effective curriculum design and assessment that is tightly woven into the curriculum itself.

2. *Instrumentation and Design.* The extent to which educators need to be concerned about the quality of assessment instruments and the approaches used to collect data is highly dependent on the underlying rationale for assessment. Highly validated instruments should always be used in research, if only because the quality of conclusions drawn from data depends strongly on the quality of the data itself; the quality of data in turn depends almost entirely on the quality of the sampling instrument and sampling design. Assessment for evaluation purposes, to understand general trends in impacts of an experience on learners, for example, can occur with lower quality instruments, although even these should be as well developed as possible. In essence, investigators need to be careful about the conclusions that they draw from data, whether those conclusions are used to assign a grade, recommend an exhibit, or publish an education research paper. In this light, educators are strongly encouraged to look deeply into the validity of the instruments they are using for assessment and to consider carefully who they are collecting data from, how they are collecting data, and how those data will be analyzed. Each of these considerations should be viewed through the lens of overarching research questions, generally with research questions aligned tightly with desired learning outcomes.

3. *Special Considerations: Cultural Validity.* The vast majority of assessments used in education are created with a hypothetical global learner in mind. Assessment considers overarching concepts, generally without consideration of the diversity of experiences and backgrounds of learners. Just as this diversity is important to include in discussions about human impacts on Earth, diversity should be considered in learner assessment. The extent to which an assessment instrument is valid for different groups of learners is a difficult issue to address, a concept that has been acknowledged and discussed extensively in science education [e.g., *Solano-Flores and Nelson-Barber*, 2001; *Lee and Luykx*, 2007; *Geraghty Ward et al.*, in press]. The social, cultural, and geographic contexts from which learners emerge will influence how information is understood and processed, and hence how well learners will perform on assessments. Aligning assessments with the learner's unique context is ideal, although generating assessments that are culturally valid is not necessarily an easy task. Educators are encouraged to balance the availability of high quality instruments with the unique contexts of their learners; where possible, assessment instruments should be adapted to align closely with the sociocultural context of the population being assessed.

4. *Special Considerations: Learners with Diverse Abilities.* As discussed, assessment in most populations, from K–12 to university to the general public, typically takes the form of tests that use verbal prompts and expect participants to either read and choose a response or read and write a response. Rarely, tests may ask for a visual response, such as a picture, although these often too require written language. Some learners may be less able to read and respond to assessments that are heavily invested in the written word. Much like the issue that different cultural contexts raises for the validity of assessment, the diverse abilities of learners may require assessments that are uniquely designed. Some learners, for example, may be too young to engage adequately with written language, may speak the language used on an assessment as a second language, or may have cognitive or other impairments that inhibit written assessment. The extent to which these learners are being engaged in education will be carefully considered by educators designing instruction, specifically because all members of society will need to be compatriots in making decisions about human–Earth interactions. In the same vein, assessment should be designed to be approachable to all learners. This may require thinking outside the box to use multiple modes of communication, including audio, video, and drawings, in implementing assessment and allowing learners to respond to assessments in similar multimodal ways [*Laing and Kamhi*, 2003].

## 5.7. AGENDA FOR FUTURE ASSESSMENT EFFORTS

Several strengths are evident in the current assessment practices that fall under the umbrella of civic understanding of the Anthropocene. Researchers have developed assessment instruments that consider the broad range of cognitive, attitudinal, and behavioral outcomes that might be expected from educational experiences available in formal and informal settings. From existing assessment data, we can build an understanding of how well people in today's world recognize human impacts on Earth. This then sets the stage for future curriculum development and educational intervention. The state of current practice and available assessment instruments also suggests several areas for future effort. In particular, we as a community must begin to engage in meaningful discussion of what is needed in assessment. Assessment needs should be considered across the broader community, rather than for individual projects, building a community of assessment practice that can share resources and be informed by findings originating in other settings.

A number of research domains, such as environmental science, educational psychology, cognitive science, and psychometrics, have traditions that guide development and use of assessment instruments. Each of these fields view assessment through slightly different lenses, recognizing different purposes for assessment and assigning different value to the assessment process and outcomes. As a consequence, different methodological foundations and norms exist across disciplines, making it difficult for tools designed in one field to be used in a second field. Encouraging collaboration across different research traditions will likely result in high-quality research instruments that are understood and valued by the many groups interested in educating people about the Earth, Earth processes, and humans' role in shaping Earth's surface and future. Collaboration also has the potential to address some of the stickier issues in assessment, such as how to accommodate diverse learners in assessment practice. Finally, multiple scholarly traditions will need to come together to ensure that assessment data are used to inform education; assessment has only truly been completed when assessment data are used to modify educational practice. This use of assessment to inform practice may require significant revamping of longstanding curriculum in both formal and informal education. Ultimately, educators must be willing to ask themselves what it means for education to be effective, what evidence will be used to assess effectiveness, and how we as educators will respond to assessments. What does it mean for education to be engaging in best practice, and how willing are we to change what we do in response to assessment data?

## REFERENCES

Aikenhead, G. S. (1985), Collective decision making in the social context of science, *Sci. Ed.*, 69(4), 453–475. doi:10.1002/sce.3730690403.

Aikenhead, G. S. (1989), Decision-making theories as tools for interpreting student behavior during a scientific inquiry simulation. *J. Res. Sci. Teaching*, 26(3), 189–203. doi:10.1002/tea.3660260302.

Alerby, E. (2000), A way of visualising children's and young people's thoughts about the environment: A study of drawings, *Envir. Ed. Res.*, 6(3), 205–222. doi:10.1080/13504620050076713.

Bowker, R. (2007), Children's perceptions and learning about tropical rainforests: An analysis of their drawings, *Envir. Ed. Res.*, 13(1), 75–96. doi:10.1080/13504620601122731.

Brown, J. M., J. Henderson, and M. P. Armstrong (1987), Children's perceptions of nuclear power stations as revealed through their drawings, *J. Environ. Psych.*, 7(3), 189–199. doi:10.1016/S0272-4944(87)80029-4.

Cheng, M., and J. K. Gilbert (2009), Towards a better utilization of diagrams in research into the use of representative levels in chemical education, in P. J. K. Gilbert and P. D. Treagust (Eds.), *Multiple Representations in Chemical Education* (pp. 55–73). Springer Netherlands, retrieved from http://link.springer.com/chapter/10.1007/978-1-4020-8872-8_4.

Connell, S., J. Fien, J. Lee, H. Sykes, and D. Yencken (1999), If it doesn't directly affect you, you don't think about it': A qualitative study of young people's environmental attitudes in two Australian cities, *Environ. Ed. Res.*, 5(1), 95–113. doi:10.1080/1350462990050106.

Cutter-Mackenzie, A., and R. Smith (2003), Ecological literacy: The "missing paradigm" in environmental education (part one), *Environ. Ed. Res.*, 9(4), 497–524. doi:10.1080/1350462032000126131.

De Jager, H., and F. van der Loo (1990), Decision making in environmental education: Notes from research in the Dutch NME-VO Project, *J. Environ. Ed.*, 22(1), 33–43. doi:10.1080/00958964.1990.9943044.

DeVellis, R. F. (2011), *Scale Development: Theory and Applications* (3rd ed.), Sage Publications, Inc., Thousand Oaks, CA.

Driver, R., P. Newton, and J. Osborne (2000), Establishing the norms of scientific argumentation in classrooms, *Sci. Ed.*, 84(3), 287–312. doi:10.1002/(SICI)1098-237X(200005)84:3<287::AID-SCE1>3.0.CO;2-A.

Dunlap, R. E., K. D. Van Liere, A. G. Mertig, and R. E. Jones (2000). New trends in measuring environmental attitudes: Measuring endorsement of the new ecological paradigm: A revised NEP scale. *J. Social Issues*, 56(3), 425–442. doi:10.1111/0022-4537.00176.

Geraghty Ward, E. M., S. Semken, and J. C. Libarkin (in press). The design of place-based, culturally informed geoscience assessment. *Journal of Geoscience Education*.

Grace, M. (2009), Developing high quality decision-making discussions about biological conservation in a normal classroom setting, *Int. J. Sci. Ed.*, 31(4), 551–570. doi:10.1080/09500690701744595.

Haladyna, T. M. (2004), *Developing and Validating Multiple-choice Test Items* (3rd ed.), Routledge, New York.

Hungerford, H., R. B. Peyton, and R. J. Wilke (1980), Goals for curriculum development in environmental education, *J. Environ. Ed.*, 11(3), 42–47. doi:10.1080/00958964.1980.9941381.

Jarrett, L. E., B. Ferry, and G. Takacs (2012), Development and validation of a concept inventory for introductory-level climate change science, *Int. J. Innov. Sci. Math. Ed.* (formerly CAL-laborate International), 20(2), retrieved November 7, 2013, from http://ojs-prod.library.usyd.edu.au/index.php/CAL/article/view/5814.

Keller, J. M. (2006), Part I. Development of a concept inventory addressing students' beliefs and reasoning difficulties regarding the greenhouse effect, Part II. Distribution of chlorine measured by the Mars Odyssey Gamma Ray Spectrometer. Proquest Dissertations And Theses 2006. Section 0009, Part 0606 446 pages; [Ph.D. dissertation]. United States – Arizona: The University of Arizona; 2006. Publication Number: AAT 3237466. Source: DAI-B 67/10, Apr 2007. Retrieved November 7, 2013, from http://adsabs.harvard.edu/abs/2006PhDT.........8K.

King, P. M., and M. J. Mayhew (2002), Moral judgment development in higher education: Insights from the defining issues test, *J. Moral Educ.*, 31(3), 247–270. doi:10.1080/0305724022000008106.

Kuhn, D., V. Shaw, and M. Felton (1997), Effects of dyadic interaction on argumentative reasoning, *Cognition Instruct.*, 15(3), 287–315. doi:10.1207/s1532690xci1503_1.

LaDue, N. D., and S. K. Clark (2012), Educator perspectives on Earth system science literacy: Challenges and priorities, *JGE*, 60(4), 372–383. doi:10.5408/11-253.1.

Laing, S. P., and A. Kamhi (2003), Alternative assessment of language and literacy in culturally and linguistically diverse populations, *Lang., Speech, Hear. Serve. Sch.*, 34(1), 44–55. doi:10.1044/0161-1461(2003/005).

Lee, O., and A. Luykx (2006), *Science Education and Student Diversity: Synthesis and Research Agenda*, Cambridge University Press, Cambridge, England.

Libarkin, J., and S. Anderson (2006), The geoscience concept inventory: Application of Rasch analysis to concept inventory development in higher education, in X. Liu and W. J. Boone (Eds.), *Applications of Rasch Measurement in Science Education*, (pp. 45–73), JAM Press, Minnesota.

Libarkin, J. C., and S. W. Jardeleza (2013), Concept inventories as research instruments, *J. Res. Sci. Teach.*, in review.

National Research Council (1996), *National Science Education Standards*, The National Academies Press, Washington, DC.

National Research Council (2012), *A Framework for K–12 Science Education: Practices, Crosscutting Concepts, and Core Ideas*, The National Academies Press, Washington, DC.

Nehm, R. H., M. Ha, and E. Mayfield (2012), Transforming biology assessment with machine learning: Automated scoring of written evolutionary explanations. *J. Sci. Education and Technology*, 21(1), 183–196. doi:10.1007/s10956-011-9300-9.

Nicolaou, C. T., K. Korfiatis, M. Evagorou, and C. Constantinou (2009), Development of decision-making skills and environmental concern through computer-based, scaffolded learning activities, *Environ. Ed. Res.*, 15(1), 39–54. doi:10.1080/13504620802567007.

Novak, J. D. (1990), Concept mapping: A useful tool for science education, *J. Res. Sci. Teach.*, 27(10), 937–949. doi:10.1002/tea.3660271003.

Patton, M. Q. (2002), *Qualitative Research and Evaluation Methods* (3rd ed.), Sage Publications, Thousand Oaks, CA.

Rodrıguez, H., W. Dıaz, J. M. Santos, and B. E. Aguirre (2007), Communicating risk and uncertainty: Science, technology, and disasters at the crossroads, in *Handbook of Disaster Research* (pp. 476–488), Springer, New York, NY, retrieved from http://link.springer.com/chapter/10.1007/978-0-387-32353-4_29

Sadler, T. D. (2004), Informal reasoning regarding socioscientific issues: A critical review of research, *J. Res. Sci. Teach.*, 41(5), 513–536. doi:10.1002/tea.20009.

Salton, G., C. S. Yang, and C. T. Yu (1975), A theory of term importance in automatic text analysis, *J. Am. Soc. Inform. Sci.*, 26(1), 33–44. doi:10.1002/asi.4630260106.

Sawada, D., M. D. Piburn, E. Judson, J. Turley, K. Falconer, R. Benford, et al. (2002), Measuring reform practices in science and mathematics classrooms: The reformed teaching observation protocol, *Sch. Sci. Math.*, 102(6), 245–253. doi:10.1111/j.1949-8594.2002.tb17883.x.

Schwarz, N. (2000), Emotion, cognition, and decision making, *Cognition Emotion*, 14(4), 433–440. doi:10.1080/026999300402745.

Solano-Flores, G., and S. Nelson-Barber (2001), On the cultural validity of science assessments, *J. Res. Sci. Teach.*, 38(5), 553–573. doi:10.1002/tea.1018.

Stapp, W. (1970), The concept of environmental education, *Ed. Digest*, 35(7), 8–10.

Volk, T. L. (1984), Project synthesis and environmental education, *Sci. Educ.*, 68(1), 23–33. doi:10.1002/sce.3730680106.

Vom Lehn, D. (2006), Embodying experience: A video-based examination of visitors' conduct and interaction in museums, *Eur. J. Marketing*, 40(11/12), 1340–1359. doi:10.1108/03090560610702849.

Wiggins, G., and J. McTighe (2001), *Understanding by Design*, Prentice Hall, Englewood Cliffs, NJ.

Wilson, V. (1997), Focus groups: A useful qualitative method for educational research? *Brit. Educ. Res. J.*, 23(2), 209–224. doi:10.1080/0141192970230207.

# 6

# Community-Driven Research in the Anthropocene

Rajul E. Pandya*

## 6.1. INTRODUCTION

The Anthropocene, as outlined in the introduction to this volume, is defined by the unprecedented global impact human society has made, and will continue to make, on the Earth system. Never before have human actions directly and indirectly impacted the lives and livelihood of ecosystems and people who were far away and yet to be born.

Our ways of doing and applying science grew up before the Anthropocene and are still adapting to this new reality. Science has already undergone two paradigm shifts in the Anthropocene: a shift away from determinism driven by the insights of quantum mechanics and chaos theory and a shift from reductionism toward systems thinking. Geoscience played a major role in both of these shifts. It will also play a major role in a third shift, as we adapt scientific methods and ideas to the challenge of doing science in an increasingly interconnected world and recognize humans as part of the Earth system.

The gap between science and society will motivate this next change. You can see evidence of this gap throughout the geosciences, in the growing socioeconomic impact of natural disasters, the politicized debates about human-induced climate change, and the difficulty in recognizing and planning for diminishing supplies of fossil fuels. It is also visible across the sciences, in low levels of public understanding of science [*National Science Board (NSB)*, 2012], students' disinterest and poor-performance in science and engineering [*NSB*, 2012], and the conflict between science and other ways of knowing, as epitomized by the longstanding controversy in the United States over teaching evolution in public schools [e.g., *Berkman and Plutzer*, 2011]. The low rates of minority participation in science [*National Science Foundation (NSF)*, 2013] in the United States suggest this science–society gap is biggest for communities that have been, and continue to be, underrepresented in science.

In the Anthropocene, the gulf between scientific understanding and civic decision-making simultaneously increases the likelihood of disaster, our vulnerability to natural hazards, and the inequity of their impact. Hurricane Katrina provides a vivid illustration. Scientists long warned about the combination of fragile physical environment and declining socioeconomic infrastructure [*Travis*, 2005; *Comfort*, 2006] that exacerbated New Orleans' risk. Nonetheless, decision-making designed to minimize the impact of frequent small events increased the vulnerability to less frequent, stronger events—a common and well-documented pattern [*Kates et al.*, 2006]. For example, landfill development of the wetlands reduced the occurrence of seasonal flooding and also eliminated a natural buffer from strong winds and storm surges [*Farber*, 1987]. Similarly, the levees were only designed to withstand a "standard project hurricane," (about a Category 3) but could be overtopped by stronger hurricanes [*Sills et al.*, 2008]. Meanwhile, anthropogenic climate change increased both the likelihood of a stronger hurricane and the strength of the associated storm surge [*McInnes et al.*, 2003]. Finally, the strong racial and class differences in the impact of the storm [*Elliott and Pias*, 2006] raises difficult questions about inequitable application of scientific research and underscores the urgency of applying science for *all* of society.

Hurricane Katrina also illustrates a fundamental point: disasters result from the combination of physical events—environmental phenomena such as drought or

---

**Thriving Earth Exchange, American Geophysical Union, Washington, D.C., and University Corporation for Atmospheric Research, Boulder, Colorado*

---

*Future Earth—Advancing Civic Understanding of the Anthropocene, Geophysical Monograph 203*, First Edition.
Edited by Diana Dalbotten, Gillian Roehrig, and Patrick Hamilton.
© 2014 American Geophysical Union. Published 2014 by John Wiley & Sons, Inc.

earthquakes—and the social, economic, political, and cultural environments that structure how people live and make them more or less vulnerable to those events [*Wisner*, 2004]. Anticipating, mitigating, and recovering from disasters, therefore, requires the integration of multiple kinds of scientific knowledge into the broader social context used to support decisions [*Alexander*, 1997]. In other words, living in the Anthropocene requires we bring science and society closer together.

Our continuing descent into the Anthropocene argues for a new approach. The large difference between the scientific consensus and public opinion about anthropogenic climate change—97 percent publishing climate scientists agree that humans' activities are contributing to a changing climate [*Anderegg*, 2010] versus only 40 percent of Americans [*Leiserowitz*, 2011]—points to a basic communication gap. The polarized nature of belief in climate change suggests that scientific evidence alone is not sufficient to affect change or impact behavior [*Moser and Dilling*, 2006] and challenges us to better integrate scientific knowledge into cultural, ethical, and aesthetic frameworks. Indeed, the notion that political opinions can influence belief in empirical phenomena is frustrating to many scientists and highlights some of the challenges of expecting scientific findings to influence actions and policy.

## 6.2. MIND THE GAP

Because this book is aimed at scientists and science educators, this chapter focuses on what scientists and science educators can do to bridge the science–society gap. To do this, the chapter begins by exploring how scientist and science educators contribute to the gap. This is not meant to blame scientists, paint them all with a broad brush, or excuse the unhelpful approaches of some non-scientists; instead it is meant to identify things scientists and science-educators could do differently that would have a positive impact.

The cultural norms, or set of expectations and rules for behavior and interaction, associated with science contribute to the gap between scientists and non-scientists. For example, the competitive norm in science shows up in introductory science classes and the focus on "weeding out" students; this in turn contributes to college students' decision to leave or avoid science majors [*Tobias and Fehrs*, 1991; *Seymour and Hewitt*, 1994, *Strenta et al.*, 1994; *Luppino and Sander*, 2012]. A corresponding devaluing of collaborative processes shows up in the tendency to value single-authored publications above multi-authored publications in tenure and promotion [*Macfarlane and Luzzadder-Beach*, 1998], despite the fact that the number of co-authored papers has grown over the last forty years in nearly every field of science [*O'Brien*, 2012], including geosciences [*Engelder*, 2007].

As another example, from personal experience, I have seen the scientific norm of skepticism (i.e., the critical scrutiny of ideas before acceptance) create tension when overused in social contexts that call for support for students or respect for elders. Communalism, or the norm that makes scientific results the common property of the entire scientific community [*Merton*, 1973], can conflict with the notion that some kinds of indigenous knowledge are privileged and only appropriate for a specific time, place, or community [*Thornburgh*, personal communication, 2009].

Even the norm of universalism [*Merton*, 1973] or the belief that anyone can make a contribution to science regardless of race, gender, or ethnicity can interfere with the connection between scientists and non-scientists. Some scientists conflate the intent of science with the practice and assume that biases and preconceptions are not active in the conduct and evaluation of science. Research, not to mention the readily apparent dearth of minorities in many sciences, demonstrates that biases do influence decisions such as hiring and mentoring [*Moss-Racusina et al.*, 2009]. This visible difference between the aspirational norm and the actual practice can undermine the overall credibility of scientists. Insidiously, the presence of the aspirational norm may exacerbate the problem by discouraging people from acknowledging and addressing bias [*Valian*, 1999].

Communication norms also contribute to the science–society gap. At the most basic level, communication to the public is often valued less than communication to other scientists. Excellent public communication may even be penalized: Carl Sagan's denial of membership in the National Academies of Science was partially attributed to his success connecting with general audiences [*Poundstone*, 1999]. More practically, the strategies scientists learn to communicate with each other may not work as well in communicating with the public. Whereas scientists focus on the content of the presentation and their argument, many non-scientists, or scientists operating outside of their own discipline, look to noncontent-related cues (such as style of dress, manner of speech, clarity of graphics) to judge the credibility of a scientific messenger and her or his message [*Olson*, 2007]. Critical questioning, common in scientific discussion, can alienate the general public [*Olson*, 2007]. The careful qualification of uncertainty can be confusing [*Bubela et al.*, 2000] or frustrating to non-scientists seeking actionable information [*Moser and Dilling*, 2006]. Worse yet, uncertainty may be deliberately exaggerated in an effort to influence public policy, as in the case of climate change [*Oreskes*, 2010].

For me, the most disturbing way in which scientists and science-educators contribute to the science–society gap is through the inclination of some of us to put science on top of the hierarchy of ways of knowing about the world. The increasingly specialized and high-tech nature of research, which raises barriers to doing science, may exacerbate this. Even within the sciences there is the oft-joked about hierarchy that places physics at the top and categorizes everything else as "mere stamp-collecting." Although that example is nearly comical, I have known geoscientists who complain about the lack of rigor in the social sciences. Misguided attempts to validate other ways of knowing scientifically, for example by "verifying" traditional ecological knowledge, are also evidence of this perceived hierarchy.

Even well-meaning attempts to place science alongside other ways of knowing may not bridge the science–society gap. Although this idea has been offered to mitigate the conflict between religion and science [*Gould*, 2004], it implies that knowledge can be pulled apart and compartmentalized, an idea that is inherently at odds with many indigenous worldviews. Albert Whitehat, a Lakota Elder, often said, "we didn't have a religion, we had a way of being," to underscore the integration and inseparability of practical, spiritual, ecological, and scientific ideas in Lakota culture. Indeed, many indigenous thinkers argue that traditional ecological knowledge is part of an integrated and complete belief system that has its own standards and practices for discovery and verification [*Deloria*, 1995; *Ford and Martinez*, 2000; *Piertti and Wildcat*, 2002]. Science has also been associated with a worldview that places humans as separate observers of natural systems [*Mayr*, 1977] and this stands in sharp contrast with many indigenous worldviews that places humans as part of the natural systems [*Deloria*, 1992; *Pierotti and Wildcat*, 1997].

All of these norms, attitudes, and practices are aspects and outgrowths of the "loading dock" model of science [*Cash et al.*, 2006]. This model was introduced in the book *Science the Endless Frontier*, which served as the blueprint for the organization and funding of academic and government research in the US after World War II [*Kelves*, 1977, p. 12]. In it, the author asserts, "the centers of basic research are the wellsprings of knowledge and understanding … [and] there will be a flow of new scientific knowledge to those that can apply it to practical problems in government, industry, or elsewhere."

By asserting that research excellence alone is sufficient to produce societal benefit, this model set the stage for the science–society gap we see today. It freed scientists from the responsibility of connecting research to practical problems and allowed decisions about science priorities to be made with minimal input from non-scientists [*Sarewitz and Pielke*, 2007]. Indeed, some have argued that the input of non-scientists would even be detrimental to the advancement of science because it would constrain free inquiry [e.g., *Polyani*, 1962].

A mild modification of the loading dock is the "science push" model of science–policy interaction, in which scientists are the primary decision makers about which projects to pursue, pursuit of knowledge is the leading criterion for setting research directions, and application to policy comes from scientists mining their findings [*Stokes*, 1977]. Even this modified loading dock does not bridge the gap between science and society. For example, much of the gap between climate science policy and climate research has been attributed to the overuse of the "scientist-push" model [*Dilling and Lemos*, 2011].

A common feature of both scientist-push model and loading-dock model is that they both begin with scientists defining the questions to pursue, as shown in the left-hand side of Figure 6.1. When the public is not included in any part of the resulting process (the left-most path in Fig. 6.1) the isolation of scientists can breed an insular culture with norms and values that diverge from the larger culture. Even if nonscientists are invited into the process later (either to contribute data or to collaborate on the analysis, as shown in Fig. 6.1) scientists remain the intellectual leaders, and this hierarchy can reinforce the tendency to elevate scientific approaches ahead of local knowledge and prioritize scientific goals over societal benefit. All the scientist-driven approaches on the left-side of Figure 6.1 ultimately lead to "push" education and application where scientists mine their results to share the results they think society might be, or should be, interested in.

## 6.3. CLOSING THE GAP

Closing the gap between science and society does not require entirely abandoning the scientist-driven approach to research. Even if the goal of research is to produce societal benefit, scientist-directed research—even research that is motivated only by curiosity—can lead to unforeseen societal benefits and should therefore stay part of the portfolio of approaches [*Leshner*, 2005]. Even outside of societal benefit, curiosity-driven scientist-led research is important simply for the value many attach to advancing human understanding of the world.

It is also worth pointing out that the simple division based on who asks the question in Figure 6.1 is overly clear-cut. Even in the last half-century while the loading-dock model has been prominent and scientists have taken the lead in asking research questions, decisions about funding research programs have long been influenced by desired societal outcomes [*Sarewitz and Pielke*, 2007].

Nevertheless, understanding and especially responding to the challenges posed by the Anthropocene requires additional approaches to science that move beyond scientists asking the questions. These approaches place more

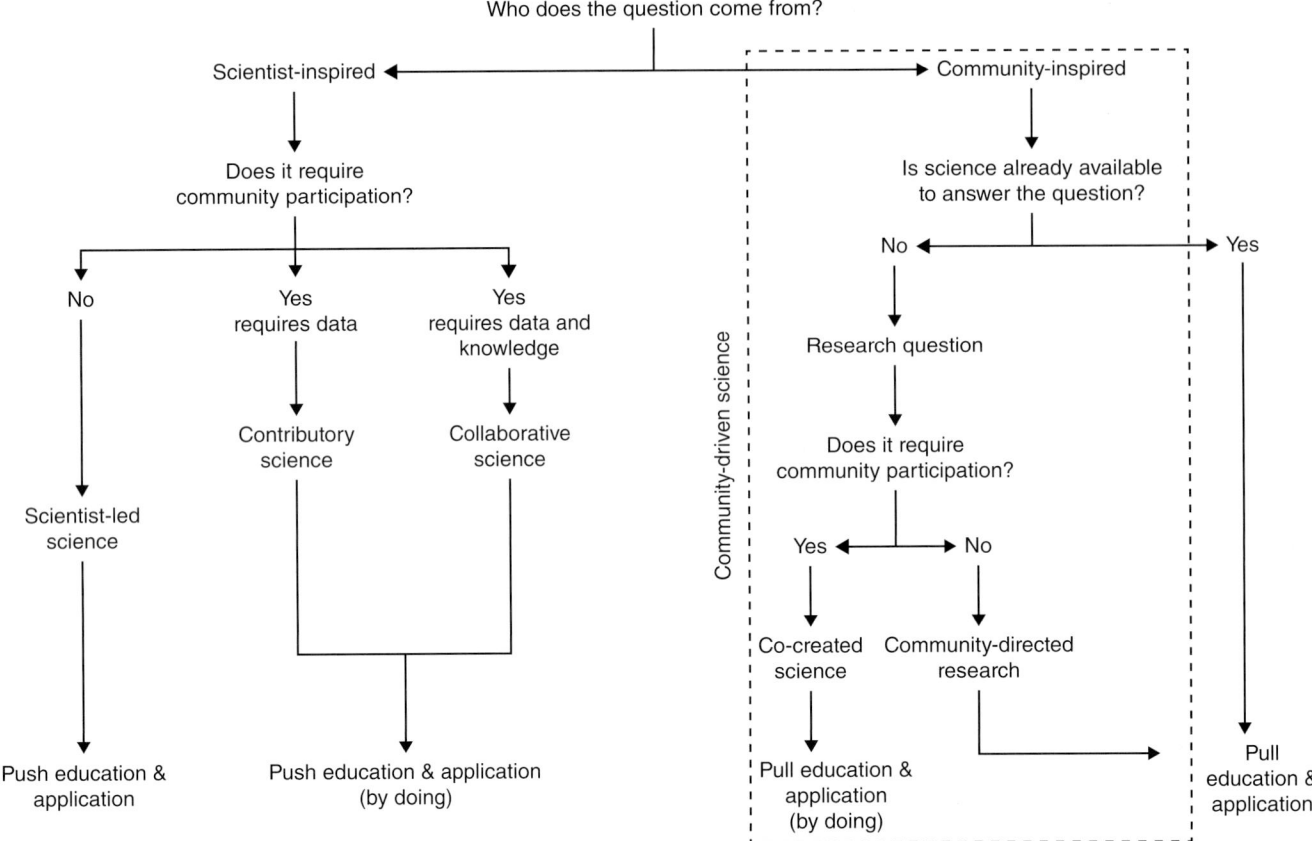

**Figure 6.1** A schematic tracing the different paths for connecting science and society. A key distinction is who participates in defining the scientific question, and this distinction flows into whether science results are pushed out from scientists or pulled into community priorities. For color detail, please see color plate section.

emphasis on the application of science, the participation of non-scientists, and the willingness to include science as one of many tools for learning about the world. At their core, all the approaches on the right side of Figure 6.1 share a commitment to inviting non-scientists to guide research priorities and define socially relevant research questions. These additional approaches will supplement the scientist-driven model and, by enhancing public benefit from science, may even increase the willingness to fund all modes of science, including the scientist-led curiosity-driven research on the extreme left side of Figure 6.1.

These new approaches expand the desired outcomes of research to include both scientific insight and usability of results. This move has been variously referred to as Jeffersonian Science [*Holton and Sonnert*, 1999], use-inspired basic research [*Stokes*, 1977], post-normal science [*Funtowicz and Ravetz*, 1993], mode 2 science [*Gibbons et al.*, 1994] and solutions-oriented science [*Crow*, 2010]. Jeffersonian and use-inspired science attempt to find a middle ground between societal and scientific priorities, by suggesting, for example, that specific projects might be situated in an area "of basic scientific ignorance that seems to lie at the heart of a social problem" [*Holton and Sonnert*, 1999]. Solutions-oriented science, mode 2 science, and post-normal science go a step farther and suggest that research can actually be initiated to address societal priorities, particularly in the context of environmental systems and sustainability. As such, these models seem particularly appropriate to the challenges of the Anthropocene.

Post-normal, mode2, and solutions-oriented science rest on a series of insights about how scientific information or insight becomes used by nonscientists. They emphasize the multidisciplinary nature of many of the questions [*Crow*, 2010]; the tight coupling of research, communication, and use [*Sarewitz and Pielke*, 2007]; iterative interaction between users and producers of scientific knowledge [*Dilling and Lemos*, 2011]; and the need for individuals or institutions to mediate between scientists and decision makers [*Gibbons et al.*, 1994].

The shared practical feature of all these models is their emphasis on including non-scientists in the process of science, and especially in decisions about which science to pursue. All these approaches, therefore, are community driven and fall on the right hand side of Figure 6.1. In all cases except the right-most path in Figure 6.1, these approaches to science also emphasize close and continual interaction between scientists and non-scientists [*Cash et al.*, 2006; *Sarewitz and Pielke*, 2007; *Dilling and Lemos*, 2011].

In examining the factors that make scientific information likely to be used in decision making, *Gibbons et al.* [1994] found that scientific findings were less likely to be contested when non-scientists were part of the scientific process. Further, interaction with a diverse collection of non-scientists—especially in setting research priorities—is necessary as a way to ensure that science is responsive to the needs of *all* who have a stake in the research outcome [*Kitcher*, 2001] and that new knowledge is not preferentially available to members of certain groups [*Bozeman and Sarewitz*, 2005]. In the words of one colleague from an underrepresented community, "if you aren't at the table, you're on the menu."

There are several related frameworks for community-driven research that span a number of fields: community-based participatory research in public health [*Israel et al.*, 1998], participatory action research in disaster management [*Park*, 1993], community-based natural resource management [*Berkes*, 2004], co-created citizen science [*Bonney et al.*, 2009], and the coproduction mode for science-policy interaction [*Dilling and Lemos*, 2011].

In public health, community-driven approaches have been motivated by the gap between research and application, the lack of research that attends to marginalized communities, and increased sensitivity to working across cultures [*Israel et al.*, 1998]. The advantages of community-driven approaches are well documented and include: better use of research results, refined research questions, enhanced research and management skills for scientists and non-scientists who participate, new employment opportunities for community members, new funding opportunities for researchers, strengthened social networks in the community, and improved relations between research institutions and their partnering communities [*Israel et al.*, 1998].

Conservation practices, especially in emerging economies, moved toward community-based resource management as a reaction to the failures of exclusionary conservation, where natural resources were thought to be best protected by isolation from humans [*Berkes*, 2004]. They have also been connected to a move toward systems-thinking in ecology [*Berkes*, 2004]. The shift to participatory conservation is linked to a better appreciation of the specific strategies indigenous people use to live sustainably [*Fabricus*, 2004] and builds on traditional ecological knowledge that includes people as active parts of the natural world [*Pierotti and Wildcat*, 2000]. Participatory approaches are shown to enhance a community's adaptive capacity or resilience [*Armitage*, 2005].

As in public health, community-driven approaches to disaster risk management were motivated by concerns about equitable benefit from disaster research especially across class and ethnicity [*Galliard et al.*, 2007]; frustration over the slow implementation of new strategies relative to the pace of research [*Glantz*, 2001]; the ever-increasing impact of disasters [*Wisner*, 2004]; and a growing recognition that social systems play as much of a role in disaster as physical systems [*Mercer et al.*, 2008]. In the context of climate change, community-driven approaches have also been motivated by a desire to ensure that poorer communities are not burdened by having the risks caused by climate change shift in their direction [*Yamin et al.*, 2005].

Citizen-science (which is being renamed as Public Participation in Scientific Research (PPSR) because the word *citizen* has become polarized by the ongoing US debate about immigration) also engages non-scientists in scientific research. Not all PPSR approaches are community driven—in contributory and collaborative models, the public is engaged in data collection or analysis but the research goals are defined by scientists (see Fig. 6.1). Co-created PPSR projects, however, involve communities in every stage of the scientific process, especially defining a scientific question, as on the right side of Figure 6.1. As in contributory and collaborative PPSR projects, nonscientists who participate gain scientific knowledge and become more comfortable with science [*Bonney et al.*, 2009], but co-created projects also enhance participants' social capital and economic opportunity and enhance the community's overall technical capacity [*Ballard and Huntsinger*, 2006; *Shirk et al.*, 2012]. In co-created projects, scientists have access to data that would otherwise go unnoticed and uncollected and learn from local insight and community knowledge. An example of a co-created PPSR project is the work of tribal college students on the White-Earth Indian Reservation to include remote sensing data in the process used to allocate permits for wild rice harvesting discussed later in this chapter [*Bennett*, personal communication, 2013].

## 6.4. COMMON ELEMENTS OF COMMUNITY-DRIVEN SCIENCE

Regardless of the field, community-driven science projects share a number of common elements and premises:

### 6.4.1. Begin with a Community-Question

Collaborative definition of a research question is the critical first step in all community-driven projects (see Fig. 6.1). This is often an exploratory and iterative process in which scientists and community members work together to identify the overlaps of community priorities and scientific capabilities. Techniques such as concept mapping, facilitated dialogues, and town-hall meetings can help refine the problem.

As the problems are defined, they may be answerable with available scientific knowledge. This is the rightmost path in Figure 6.1, and it encompasses participatory approaches to education [e.g., *Friere*, 2000]. Community-driven research, however, requires community questions that push at the boundaries of what is known scientifically. Some of these questions might be answered by scientists working in isolation from communities (community-directed science in Fig. 6.1), but many of these questions will require community participation to access local knowledge and theory based on the lived experience of the people involved. This is the right-most and most participatory path under community-driven science, co-created science.

### 6.4.2. Embrace Multiple Priorities

Community-driven participatory research is built on the idea of mutual benefit to all parties, including communities [*Israal et al.*, 1998]. Because community goals go beyond simply contributing to scientific knowledge, participatory research must address a host of goals, including addressing or managing specific environmental challenges [*Hunnington*, 2000; *Probst et al.*, 2003], enhancing economic growth and opportunity, increasing community-members technical skills [*Viswananathan et al.*, 2004], informing and supporting community-led advocacy [*Park*, 1993], and enhancing social ties in the community [*Heany and Israel*, 2002]. For example, one successful strategy that has been used to enhance the diversity of participants in citizen science programs is to attach the citizen-science program to existing programs whose primary goals are youth empowerment or community clean-up [*Porticella et al.*, 2013]. Similarly, the White Earth Nation's investigation of wild rice's future grew out of wild rice's central importance to the cultural education of tribal members, its contribution to family-level food and economic security, the opportunities wild rice provides for tribal economic growth from wild rice export, and the contribution wild rice can make to healthy diets [*Bennett*, personal communication, 2013].

### 6.4.3. Value Community Knowledge

Successful participatory projects seek expertise from all participants and agree to processes and procedures that validate multiple kinds of expertise [*Israel et al.*, 1998]. Many successful projects value traditional and local knowledge, historical accounts, and participant observations in addition to scientific data [*Huntington*, 2000]. It is worth emphasizing that community knowledge is not limited to indigenous populations [*Huntington*, 2000]. For example, migrant and immigrant harvesters of non-timber forest products in the US Northwest [*Ballard and Huntsinger*, 2006] and people who fish for a living in the Louisiana Bayou [*Button and Peterson*, 2009] have been part of participatory science projects.

It is also important to acknowledge that community knowledge need not be confined to the realm of geosciences. Indeed, part of the point of community-based science is situating the science in a complete context that includes political, social, and ethical considerations. For example, examining the social conditions associated with heat wave mortality led to the realization that rich social networks and strong family connections offer protection from heat [*Harlan et al.*, 2006].

If community knowledge is included, it is important that all participants need to agree on what constitutes data and how those data will be collected, validated, and shared within and even beyond the project [*Huntington*, 2000]. Special care needs to be given to knowledge that is sacred or culturally sensitive. Many communities have guidelines that define the terms and conditions of their participation in research, and these should be part of the discussion between scientists and community. For communities that do not have these guidelines, the creation of such guidelines should be a part of the overall project design [*Minkler and Wallerstein*, 2010].

### 6.4.4. Iterate

Although all community-driven projects begin with a shared or even community-posed question, participatory or co-created projects engage scientists and non-scientists in all subsequent stages of research including collecting data, analyzing data, sharing results in scientific and non-scientific forums, and applying results. This may encompass extensive training and even employment opportunities [*Fazey et al.*, 2010], and it helps ensure co-ownership of the project and the application of results [*Israel et al.*, 1998]. It also contributes to better research outcomes [*Bang et al.*, 2007].

Successful participatory projects build processes and procedures (such as regular community meetings, advisory boards, frequent informal interaction between researchers and community members, community risk and asset mapping, focus groups) to plan for and encourage interaction between scientists and communities [*Israel et al.*, 1998; *Minkler and Wallerstein*, 2010]. Institutions and actors that "own" the task of creating the conditions and mechanisms for this are essential [*Dilling and Lemos*, 2010].

The most successful projects extend participatory approaches to the dissemination of results not only in the form of scientific publications but in ways that are designed to be relevant to community priorities and allow new knowledge to be easily applied [*Eden*, 2006] for all partners, in appropriate language and venues, and with ownership acknowledged [*Israel et al.*, 1998]. As with all tasks, dissemination of findings should be the shared responsibility of all project participants. In particular, community participation in scientific presentations has a positive impact on the overall relationship between the scientists and the community [*Button and Peterson*, 2009].

### 6.4.5. Leverage Diversity

Successful participatory approaches ensure equitable participation for all parts of the community and prevent the science process from enhancing or exacerbating social inequity [*Israel et al.*, 1988]. Strategies for including the whole community include broad stakeholder meetings, focused meetings with members of otherwise marginalized groups, and private conversations in neutral settings [*Peterson and Button*, 2009]. For example, some organizations routinely hold women-only meetings in parts of the world where women are discouraged from speaking up in the presence of men [*Guibert*, personal communication, 2009]. Another strategy involves actively helping community members gain political access and visibility. For example, a project in the Solomon Islands invited project participants who did not have leadership positions to take the lead in briefing government officials about the research [*Frazey et al.*, 2000].

The most successful projects also leverage diverse skills from participating scientists. Many community challenges do not map neatly to a single scientific discipline, and so the most successful teams will include expertise from multiple disciplines [*Crow*, 2010]. Participatory approaches require careful trust building between scientists and community members [*Peterson and Button*, 2009], and this opens the door to valuing and rewarding new kinds of skills in the scientific community. For example, scientists who come from diverse or underserved communities may be able to provide insight into the challenges communities face and can help other scientists learn about unfamiliar customs and practices.

### 6.4.6. Learn Together

If everyone, scientist or non-scientist, is a valued partner with knowledge to contribute, than it stands to reason that everyone must also have something to learn. Some of this learning is preparatory. Because participatory approaches are new to many scientists, formal training may be beneficial [*Button and Peterson*, 2009]. Cultural orientation can help scientists understand history, learn new approaches, and avoid mistakes. More generally, cultural competence [*Lynch and Hanson*, 2004] and cultural humility [*Tervalon and Murray-Garcia*, 1998] can be learned and can enhance the ability to interact effectively with people. The fastest and most critical learning, however, occurs when scientists are supported with mentoring and given time and resources to engage in reflection and analysis in the course of their participatory work with communities [*Button and Peterson*, 2009].

Similarly, non-scientists from the community have ample opportunities to learn from their collaboration with scientists. Indeed, one of the advantages of participatory approaches is their emphasis on the kinds of hands-on, authentic investigation that is consistent with recommendations for both formal [*Brown and Cocking*, 2000] and informal education [*Bell et al.*, 2009]. In the concept of the Anthropocene, where many civic decisions need to be made in the context of evolving scientific understanding and in the face of uncertainty, participatory approaches are important because they allow people to better apprehend these issues through firsthand experiences in the processes of science [*Brown and Cocking*, 2000]. The combination of education, training, and employment also provides a way to engage whole families, which is often a key priority for many communities [*Porticella et al.*, 2013].

## 6.5. A FEW EXAMPLES

### 6.5.1. Meningitis in the Sahel

At its broadest level, this research was motivated by the impact meningitis has in the Sahel. An epidemic in 1998, for example, resulted in 250,000 cases with an estimated 25,000 deaths and 50,000 people left permanently disabled [*World Health Organization*, 2003]. The research question at the heart of the project-"how are meningitis epidemics impacted by weather and climate?"-came directly from people in the Sahel, who have long known that meningitis is a disease of the dry, dusty season and ends with the onset of the monsoon. In fact, in some regions, meningitis is referred to colloquially as "sand disease."

Although this insight defined the broad outlines of the research, iterative interaction with public health practitioners helped focus the research question. Until recently, the protection provided by the only available vaccine was so limited and short-lived that the only practical strategy for health officials was reactive: wait until an epidemic occurs in a region, then vaccinate in and around that region to prevent the epidemic's further spread. Even with this conservative strategy, demand outpaces available vaccine. The research, then, focused on improving and tailoring weather forecasts so that public health officials can know where changing meteorological conditions will end the epidemics naturally and deploy their limited vaccines to other regions experiencing epidemics.

There are several points where this project illustrates the characteristics of community-driven research. Although the general project was generated in response to community input and priorities from people throughout the Sahel, the core participating community was the community of public health practitioners who work in the Sahel. We interacted with local as well as international public health officials in line with the emphasis on engaging across the community. To develop an economic, pragmatic, and culturally appropriate solution, the project included epidemiologists, meteorologists, anthropologists, and economists. Data collection and analysis were shared efforts—with public health practitioners bringing epidemiological data and local knowledge and meteorologists contributing environmental data. The analysis and publications were prepared iteratively, through face-to-face meetings and e-mail. Usability of the final decision–relevant product was refined through our participation in weekly phone calls hosted to make actual decisions about vaccine deployment. The project included lots of opportunities for co-learning including frequent trips to the Sahel for the international research team, project-related graduate school opportunities for partners in Africa, and frequent co-presentation at both geophysical and public-health oriented meetings. Finally, an international organization, meningitis and environmental risk information technology (MERIT) owned the overarching goal of fostering regular interaction between public health and geoscience communities and provided the framework for the interactions in this project. In terms of the different strategies for community-driven science shown in Figure 6.1, this project traced the path from a question asked by the community, to education and application that is "pull" rather than push driven; it is the co-created science or leftmost path of the community-driven paths.

### 6.5.2. Land Loss in the Louisiana Bayou

Southern Louisiana is undergoing rapid land loss [*Britsch and Dunbar*, 1993]. Natural geologic-scale subsidence is no longer offset by deposition from an active river outlet, and logging of cypress, dredging of canals to support oil exploration, the introduction of non-native species that consume local vegetation, and rising sea-levels associated with climate-change all contribute to the enhanced rate of land loss [*Day et al.*, 2000].

In 2012, to learn and practice participatory approaches to science, two undergraduate interns spent the summer in Southern Louisiana to talk with local communities about the area and its challenges and opportunities. One of the deliverables the students and community agreed to create together was an iPhone app, called Vanishing Points™ that would allow community members to locate culturally important places, collect stories and images of those places, link to projections about the places' future, and finally link to organizations or community resources that are actively working on land-loss issues.

Students working at the Southern Louisiana Wetlands Discovery Center are collecting community locations and associated stories, the National Center for Atmospheric Research's (NCAR) education and outreach group is designing the overall app, and scientists are contributing projections for the places' futures. In terms of the participatory pathways in Figure 6.1, this project is community-driven, with a co-defined problem that aligns scientific questions with the local priorities of educational opportunity, community advocacy, and community planning. The high school–student collection of stories traces the path of community-driven education (the rightmost path in Fig 6.1) and the scientists producing future projections are tracing the middle path of community-directed research in Figure 6.1.

Student participation, especially for the summer interns, made co-learning an organizing premise of this project. The two science interns had extensive mentoring on participatory methods before and during the summer, opportunities for reflection and analysis through blogs and structured conversations, and they were paired with a slightly older community member who could provide an introduction to the community.

### 6.5.3. Wild Rice and White Earth

Manoomin or wild rice is of tremendous cultural and economic importance to the Anishinaabe people of the Great Lakes region, including those who are part of the White Earth Nation. In recent years, many ricing families on White Earth and the White Earth Natural Resources Department have reported decreased rice production as a result of diminished water availability, fluctuating water levels on the rice beds, and the encroachment of invasive species [*Bennett*, personal communication, 2013]. In the future, continued biome shifts related to climate change will also impact wild rice.

White Earth Tribal and Community College is leading a comprehensive approach that exemplifies the co-creation pathway in Figure 6.1. The goals include developing strategies for managing wild rice and planning for climate change while growing tribal capacity to lead and apply scientific research. They are inviting scientists to help define and clarify the research questions, identify and access relevant data, and improve monitoring, all with the goal of increasing the use of research and data in decision making. Consistent with the college's mandate to serve the community in an open and transparent way, they are reaching into the broad tribal community to engage students and elders in the project. They are integrating many kinds of data, including local knowledge, and developing strategies for data collection, validation, and management that respect local traditions and value local knowledge. Because the college is closely connected to tribal leadership, there is a regular and structured way for the community to contribute to project management and be involved in all stages of the research [*Pandya*, 2012].

White Earth is just one of many indigenous communities that have been observing, experiencing, mitigating, and adapting to climate change. Recognizing this, nearly 50 authors have came together to create a special issue edition for the journal *Climatic Change*, "Climate Change and Indigenous Peoples in the United States: Impacts, Experiences and Actions." The appendix to this chapter discusses this special issue. The special issue as a whole reiterates the importance of viewing people's experiences of the Anthropocene in a larger social and historical context and catalogues the positive outcomes that can come from granting traditional ecological knowledges and local tribal observations the same respect as more conventionally-practiced scientific methods.

## 6.6. CONCLUSION

To address the challenges of the Anthropocene, humans need to integrate scientific knowledge into their ways of thinking about the world and making decisions. An effective way to do that is to add participatory approaches to the portfolio of scientific methods. The most engaging of these approaches is community-driven science: developing and answering questions that are driven by the needs and priorities of specific, local, diverse nonscientific communities. Community-driven science includes both practical strategies and a shift toward a more inclusive worldview that places science alongside, rather than above, other ways of knowing. In short, as scientists and educators, we need to do science *with* people, not *for* them or *at* them.

These ideas are not new; they have a long history in public health, disaster management, and citizen science, and there are models emerging in the geosciences. It is worth offering a few suggestions to the scientific community, borrowed from the long experience in public health [*Israel et al.*, 2001] to facilitate these kinds of participatory approaches.

*Funding*: participatory methods depend on building relationships between communities and researchers, and this requires more time than is available in a typical grant application—exploratory or planning grants aimed at fostering these relationships would be helpful. For communities to participate in research as equals, it would be helpful to allocate shared fiduciary responsibility for the work, and so create mechanisms that allow scientific funding to flow jointly to community organizations and researchers. Finally, participatory approaches may require a commitment to long-term projects and to supporting infrastructure, both of which may extend beyond the typical three- to four-year lifetimes of many grants.

*Review process*: The model of peer review is well-suited to judging the scientific merit of a proposed project, but it is difficult to understand how community-driven questions can be adequately scrutinized without the participation of community members and experts in participatory methods. It would be beneficial, therefore, to include non-scientists in the review of participatory proposals.

*Education*: Although co-learning is an important part of the participatory process, there should be opportunities to learn before engaging in a project. For community members, scientific, technical, and management skills in advance can help them position themselves as equals when working with researchers. Similarly, advance exposure to participatory methods and cross-cultural communication can help scientists be more effective in developing projects with communities. Educational opportunities for members of historically underserved or underrepresented communities contribute to both goals; they can enlarge the pool of researchers interested in addressing community issues and augment the capacity of the communities they come from.

*Tenure and promotion*: One barrier to participatory methods is the perceived and/or real career risks for researchers. We need to develop standards for evaluating participatory research, ways for journals to include participatory research results, and incentives that recognize and reward contributions to community goals.

## 6.7. EPILOGUE

It is easy to see the Anthropocene in terms of unprecedented, largely destructive, and abstract human impacts on the natural system. A series of artworks by Chris Jordan explored ways of making this impact visceral: expansive panoramas that look like natural systems from afar but, as you zoom in, reveal themselves to be built from lots of little pieces of human garbage. A bamboo forest, for example, turns out to be stacks and stacks of the 1.4 million paper bags used every hour in supermarkets in the United States.

Jordan's art also explores something more optimistic. A giant mandala titled "E Pluribus Unum" resolves, on closer inspection, into the names of one million organizations devoted to peace, environmental stewardship, social justice, and the preservation of diverse and indigenous culture. Like "E Pluribus Unum" and the indigenous view that makes humans part of nature, the scientific notion of the Anthropocene offers a way to see all humans as partners in building a sustainable future. Participatory community-driven science offers a strategy to engage them.

# APPENDIX
## Climate Change and Indigenous Peoples in the United States: Impacts, Experiences and Actions: Highlights from a Special Issue of *Climatic Change*

**Julie Koppel Maldonado, Benedict Colombi, Rajul Pandya, Kathy Lynn, and Dan Wildcat**

### INTRODUCTION

Indigenous ecological knowledge systems present a context for understanding ecological change and adaptive strategies for coping with such change that could provide crucial insight for indigenous communities and all people around the world [*Hardison and Williams*]. However, as the impacts of climate and other human-induced changes they are experiencing everyday become more rapid and severe, this significant knowledge among indigenous peoples is at risk of being lost. The issues currently experienced by indigenous communities in the United States as a result of climate change include: loss of traditional knowledge; forests and ecosystems; food security and traditional foods; water; Arctic sea ice loss; permafrost thaw; and relocation.

Recognizing this, those involved with the tribal chapter of the Draft Third National Climate Assessment called for a careful and respectful summary of indigenous observations, experiences, and adaptive strategies to climate change by indigenous peoples around the United States. Nearly 50 authors representing tribal communities, academia, government agencies, and nongovernmental organizations came together to create a special issue edition for the journal *Climatic Change*, "Climate Change and Indigenous Peoples in the United States: Impacts, Experiences and Actions."

Indigenous peoples maintain tribally specific, place-based cultures, sustainable livelihoods, and knowledge systems tied to culturally modified, ancestral homelands. Many tribal communities are displaced and relegated to marginal lands without access to basic resources, subject to treaties and reserved rights that designate their territories, and constrained by restrictive reservation boundaries—all of which hinder their ability to migrate or access resources as they once did if resources became scarce [*Lynn et al.*]. As the animal and plant species they interact with for their livelihoods and cultural practices move, they can no longer move with them.

Indigenous peoples continue to confront disparate levels of poverty and vulnerability, as well as political and social marginalization from centuries of oppression resulting from the colonial encounter. Climate change impacts intensify related threats and stressors already increasing within many indigenous communities. Thus, situating contemporary indigenous experiences of human-induced change necessitates a greater understanding of the socio-historical context. For example, the severe impacts experienced from European colonization on the quality and quantity of berries that the Wabanaki people of Maine and Canada rely on for their subsistence, culture, healing, and traditional practices are becoming more intense because of climate and other rapid socioecological changes [*Lynn et al.*]. *Reo and Parker* suggest that the severe impacts on coupled human-natural systems that occurred in New England where European colonization led to drastic social and environmental transformations could provide insight to today's context of rapid change. They depict how integrating colonial history and ecology are useful to help determine current, significant human-environment interactions, and adaptive strategies between tribal nations, policymakers, and the global community.

Tribal communities' vulnerability to climate change and other human-induced impacts are not simple linear problems nor are they solely physical ones; rather, these impacts threaten multigenerational tribal epistemologies and cultural value systems, which shape contemporary indigenous practices and identities. For example, the Pyramid Lake Paiute tribe are faced with significant cultural and environmental risks as the lake they economically and culturally depend on is impacted by a reduction in water from the diversions created by dams and shortages of water from sustained drought and climate change [*Gautum et al.*].

Similarly, migratory salmon in the Pacific Northwest are integral parts of tribal subsistence and cultural and spiritual livelihoods. Salmon-dependent resources are increasingly under threat as a result of climate and human-induced changes to the watershed, and so too are tribal livelihoods and a salmon-based way of life

[*Dittmer*; *Grah and Beaulieu*]. Tribal communities throughout the United States are facing similar water-resource hazards, such as risks to water quality and quantity, as a result of significant human-induced changes, including climate change [*Cozzetto et al.*].

To respectfully and effectively address these issues, tribal nations implement policies encouraging federal and state agencies and western scientists to include traditional ecological knowledge in developing adaptation plans for water resource-related impacts [*Cozzetto et al.*; *Grah and Beaulieu*]. For example, Columbia Basin tribes are engaged in intergovernmental and intertribal cooperation, resulting in the Columbia River Inter-Tribal Fish Commission and other collaborations with federal agencies such as the National Oceanographic and Atmospheric Administration and the National Fish and Wildlife Service. In salmon restoration these partnerships are effective in co-managing hatchery programs and in developing long-range management strategies [*Dittmer*].

These partnerships should be based on a just system of responsibilities [*Whyte*]. Creating a governance framework guided by the principles of justice and human rights establishes equitable support to communities facing severe consequences of climate and other human-induced changes, such as forced displacement [*Maldonado et al.*; *Whyte*]. Such a framework would support communities in places such as Alaska and coastal Louisiana leading their own relocation efforts to decrease the social, cultural, and economic impoverishment risks associated with forced displacement [*Maldonado et al.*].

In establishing greater self-governance mechanisms and partnerships, it is particularly important to pay attention to what is happening at the local level [*Doyle et al.*]. For example, based on local observations of climate-related health and water issues by elders of the Crow Tribe in Montana, the tribe and its tribal college partnered with state academic institutions to examine data from National Oceanic and Atmospheric Administration's (NOAA) local weather station. The data and local observations were brought together to develop mitigation strategies to reduce waterborne microbial health risks [*Doyle et al.*].

Despite the layers of vulnerability confronting indigenous peoples [*Gautum et al.*], they continue to use traditional knowledge systems, local observations and experiences, skills, and agency to actively adapt to climate and other anthropogenic changes. Pairing traditional ecological knowledge with western science can enhance understanding and offer new adaptation strategies. For example, the North Pacific Landscape Conservation Cooperative has collected tribal input to prioritize tribal responses and adaptation to the climate-related changes within their forests and ecosystems [*Voggesser et al.*].

The active collaborations taking place between tribes and academic institutions, as well as governmental agencies and nongovernmental organizations give the Crow tribe and Columbia River Inter-Tribal Fish Commission, for example, a base to demonstrate the value of including traditional ecological knowledge and indigenous science in climate change research and adaptation strategies. More of these multipronged approaches need to be implemented to include and mutually respect both traditional ecological knowledge and western science, bringing together data collection, local observations, experiences, and human-environmental relationships and interactions [*Cochran et al.*]. These approaches must also consider culturally sensitive tribal information and protect tribal traditional ecological knowledge [*Hardison and Williams*].

The collaborative effort between both tribal nations and nontribal representatives to create the special issue and the case studies highlighted in the enclosed articles show that indigenous peoples are shaping actions that address the many challenges they face and emphasize the importance of people coming together in strategic and respectful partnerships. Equitable and meaningful results are achieved when traditional ecological knowledge and local tribal observations are held up with the same respect as western science and when indigenous peoples who are experiencing these impacts guide the research, mitigation, and adaptation plans through what Daniel Wildcat calls indigenous ingenuity or "indigenuity." The case studies provide an inclusive view of how indigenous peoples throughout the United States are observing, experiencing, mitigating, and adapting to climate change, which have relevancy for indigenous peoples and communities facing parallel circumstances worldwide.

## Contents

Maldonado, Julie Koppel, Rajul E. Pandya, and Benedict J. Colombi, eds. (2013) Climate Change and Indigenous Peoples in the United States: Impacts, Experiences and Actions. Special Issue, *Climatic Change*. Vol 120, Issue 3.

1. Introduction: Climate Change and Indigenous Peoples in the United States
   Daniel Wildcat
   DOI: 10.1007/s10584-013-0849-6

2. Justice Forward: Tribes, Climate Adaptation and Responsibility
   Kyle Powys Whyte
   DOI: 10.1007/s10584-013-0743-2

3. Culture, Law, Risk and Governance: The Ecology of Traditional Knowledge in Climate Change Adaptation
Preston Hardison and Terry Williams
DOI: 10.1007/s10584-013-0850-0

4. The Impacts of Climate Change on Tribal Traditional Foods
Kathy Lynn, John Daigle, Jennifer Hoffman, Frank K. Lake, Natalie Michelle, Darren Ranco, Carson Viles, Garrit Voggesser, and Paul Williams
DOI: 10.1007/s10584-013-0736-1

5. Indigenous Frameworks for Observing and Responding to Climate Change in Alaska
Patricia Cochran, Orville H Huntington, Caleb Pungowiyi, Stanley Tom, F Stuart Chapin III, Henry P Huntington, Nancy G Maynard, Sarah F Trainor
DOI: 10.1007/s10584-013-0735-2

6. Climate Change Impacts on the Water Resources of American Indians and Alaska Natives in the U.S.
Karen Cozzetto, K. Chief, K. Dittmer, M. Brubaker, R. Gough, K. Souza, F. Ettawageshik, S. Wotkyns, S. Opitz-Stapleton, S. Duren, and P. Chavan
DOI: 10.1007/s10584-013-0852-y

7. Climate Change in Arid Lands and Native American Socioeconomic Vulnerability: The Case of the Pyramid Lake Paiute Tribe
Mahesh Gautam, Karletta Chief, and William J. Smith Jr.
DOI: 10.1007/s10584-013-0737-0

8. The Impact of Climate Change on Tribal Communities in the US: Displacement, Relocation, and Human Rights
Julie Koppel Maldonado, Christine Shearer, Robin Bronen, Kristina Peterson, and Heather Lazrus
DOI: 10.1007/s10584-013-0746-z

9. Cultural Impacts to Tribes from Climate Change Influences on Forests
Garrit Voggesser, Kathy Lynn, John Daigle, Frank K. Lake, and Darren Ranco
DOI: 10.1007/s10584-013-0733-4

10. Changing Streamflow on Columbia Basin Tribal Lands- Climate Change and Salmon
Kyle Dittmer
DOI: 10.1007/s10584-013-0745-0

11. Exploring Effects of Climate Change on Northern Plains American Indian Health
John Doyle, Margaret Hiza Redsteer, and Margaret Eggers
DOI: 10.1007/s10584-013-0799-z

12. The Effect of Climate Change on Glacier Ablation and Baseflow Support in the Nooksack River Basin and Implications on Pacific Salmon Species Protection and Recovery
Oliver Grah and Jezra Beaulieu
DOI: 10.1007/s10584-013-0747-y

13. Re-thinking colonialism to prepare for the impacts of rapid environmental change
Nicholas Reo and Angela Parker
DOI: 10.1007/s10584-013-0783-7

## REFERENCES

Alexander, D. (1997), The study of natural disasters, 1977–97: Some reflections on a changing field of knowledge, *Disasters*, 21(4), 284–304.

Anderegg, W. R. L, J. W. Prall, J. Harold, and S. H. Schneider (2010), Expert credibility in climate change, *Proc. Natl. Acad. Sci.*, 107(27), 12107–12109.

Armitage, D. (2005), Adaptive capacity and community-based natural resource management, *Envir. Mgmt.*, 35(6), 703–715.

Ballard, H. L., and L. Huntsinger, L. (2006), Salal harvester local ecological knowledge, harvest practices and understory management on the Olympic Peninsula, Washington, *Human Ecol.*, 34(4), 529–547.

Bang M., D. L. Medin, and S. Atran (2007), Cultural mosaics and mental models of nature, *Proc. Natl. Acad. Sci.*, 104, 13868–13874.

Bell, P., B. Lewenstein, A. W. Shouse, and M. A. Feder (Eds.) (2009), *Learning Science in Informal Environments: People, Places, and Pursuits*, National Academies Press, Washington, DC.

Berkes, F. (2004), Rethinking community-based conservation, *Conservation Biol.*, 18(3), 621–630.

Berkman, M. B., and E. Plutzer (2011), Defeating creationism in the courtroom, but not in the classroom, *Science*, 331 (6016), 404–405. DOI:10.1126/science.1198902.

Bonney, R., H. Ballard, R. Jordan, E. McCallie, T. Phillips, J. Shirk, et al. (2009), *Public Participation in Scientific Research: Defining the Field and Assessing Its Potential for Informal Science Education*, Center for Advancement of Informal Science Education (CAISE), Washington, DC.

Bozeman, B., and D. Sarewitz (2005), Public value failures and science policy, *Sci. Pub. Pol.*, 32(3), 119–136.

Britsch, L. D., and J. B. Dunbar (1993), Land loss rates: Louisiana coastal plain, *J. Coastal Res.*, 324–338.

Brown, A. L., and R. R. Cocking (2000), *How People Learn*, John D. Bransford (ed.), National Academy Press, Washington, DC.

Bubela, T., M. C Nisbet, R. Borchelt, F. Brunger, C. Critchley, E. Einsiedel, et al. (2000), Science communication reconsidered, *Ecol. Applic.*, 10(5), 1333–1340.

Button, G. V., and K. Peterson (2009), *Participatory Action Research: Community Partnership with Social and Physical Scientists in Anthropology and Climate Change*, Left Coast Press, Walnut Creek, CA.

Cash, D. W., J. C. Borck, and A. G. Patt (2006), Countering the loading-dock approach to linking science and decision making comparative analysis of El Niño/Southern Oscillation (ENSO) Forecasting Systems, *Sci. Tech. Human Values*, 31(4), 465–494.

Comfort, L. K. (2006), Cities at Risk Hurricane Katrina and the Drowning of New Orleans, *Urban Aff. Rev.*, 41.4, 501–516.

Crow, M. M. (2010), Organizing teaching and research to address the grand challenges of sustainable development, *BioScience*, 60(7), 488–489.

Day, J. W., L. D. Britsch, S. R. Hawes, G. P. Shaffer, D. J. Reed, and D. Cahoon (2000), Pattern and process of land loss in the Mississippi Delta: A spatial and temporal analysis of wetland habitat change, *Estuaries*, 23(4), 425–438.

Deloria, V. Jr. (1992), *God Is Red: A Native View of Religion*, North American Press, Golden, CO.

Deloria, V., Jr. (1995), *Red Earth, White Lies*, Harper and Row, San Francisco, CA.

Dilling, L., and M. C. Lemos (2011), Creating usable science: Opportunities and constraints for climate knowledge use and their implications for science policy, *Global Environ. Change*, 21(2), 680–689.

Eden S. (2006), Public participation in environmental policy: Considering scientific, counter-scientific and non-scientific contributions, *Public Underst. Sci.* 5, 183–204.

Elliott, J. R., and J. Pais (2006), Race, class, and Hurricane Katrina: Social differences in human responses to disaster, *Soc, Sci. Res.*, 35(2), 295–321.

Engelder, T. (2007), The coupling between devaluation of writing in scientific authorship and inflation of citation indices, *GSA Today*, July 2007, 44–45.

Fabricius, C. (2004), *The fundamentals of community-based natural resource management in rights, resources and rural development: Community-based natural resource management in Southern Africa*, Earthscan Publishing, London, United Kingdom.

Farber, S. (1987), The value of coastal wetlands for protection of property against hurricane wind damage, *J. Environ. Econ. Mgmt.*, 14(2), 143–151.

Fazey I., M. Kesby, A. Evely, I. Latham, D. Wagatora, J.-E. Hagasua, et al. (2010), A three-tiered approach to participatory vulnerability assessment in the Solomon Islands, *Global Environ. Chang.*, 20, 713–728.

Ford, J., and D. Martinez (2000), Traditional ecological knowledge, ecosystem science, and environmental management, *Ecol. App.*, 10(5), 1249–1250.

Freire, P. (2000), *Pedagogy of the Oppressed*, Continuum International Publishing Group/Bloomsbury, London.

Funtowicz, S. O., and J. R. Ravetz (1993), Science for the post-normal age, *Futures*, 25(7), 739–755.

Gaillard J. C., C. Liamzon, and J. D. Villanueva (2007), 'Natural' disaster? A retrospect into the causes of the late-2004 typhoon disaster in Eastern Luzon, Philippines, *Environ. Hazards*, 7, 257–270.

Gibbons, M., C. Limoges, H. Nowotny, S. Schwartzman, P. Scott, and M. Trow (1994), *The New Production of Knowledge: The Dynamics of Science and Research in Contemporary Societies*, SAGE Publications Limited, Thousand Oaks, CA.

Glantz, M. H. (2001), *Once Burned, Twice Shy: Lessons Learned from the 1997–98 El Niño*, United Nations University Press, Tokyo, Japan.

Gould, S. J. (2004), *The Hedgehog, the Fox, and the Magister's Pox: Mending the Gap between Science and the Humanities*, Three Rivers Press, New York, NY.

Harlan, S. L., A. J. Brazel, L. Prashad, W. L. Stefanov, and L. Larsen (2006), Neighborhood microclimates and vulnerability to heat stress, *Soc. Sci. Med.*, 63(11), 2847–2863.

Heaney, C. A., and B. A. Israel (2002) Social networks and social support, *Health Behav. Health Ed., Theor. Res. Pract.*, 3, 185–209.

Holton, G., and G. Sonnert (1999), A vision of Jeffersonian science, *Iss. Sci. Tech.*, 16(1), 61–65.

Huntington H. P. (2000), Using traditional ecological knowledge in science: methods and applications, *Ecol. Appl.*, 10, 1270–1274.

Israel, B. A., A. J. Schulz, E. A. Parker, and A. B. Becker (1998), Review of community-based research: Assessing partnership approaches to improve public health, *Ann. Rev. Publ. Health*, 9, 173–202.

Israel, B. A., A. J. Schulz, E. A. Parker, and A. B. Becker (2001), Community-based participatory research: policy recommendations for promoting a partnership approach in health research, *Educ. Health*, 14(2), 182–197.

Kates, R. W., C. E. Colten, S. Laska, and S. P. Leatherman (2006), Reconstruction of New Orleans after Hurricane Katrina: A research perspective, *Proc. Nat. Acad. Sci.*, 103(40), 14653–14660.

Kevles, D. J. (1977), The National Science Foundation and the debate over postwar research policy, 1942–1945: A political interpretation of science—The endless frontier, Isis, 68(1), 5–26.

Kitcher, P. (2001), *Science: Truth and Democracy*, Oxford University Press, Oxford, United Kingdom.

Leiserowitz, A., E. Maibach, C. Roser-Renouf, and N. Smith (2011), *Yale Project on Climate Change Communication: Global Warming's Six Americas in May 2011*, Yale University and George Mason University, New Haven, CT.

Leshner, A. I. (2005), Where Science Meets Society, *Science*, 309 (221), 8.

Luppino, M., and R. Sander (2012), College Major Competitiveness and Attrition from the Sciences. Available at Social Science Research Network eLibrary (SSRN), www.ssrn.com, 2167961.

Lynch, E. W., and M. J. Hanson (2004), *Developing Cross-Cultural Competence: A Guide for Working with Children and their Families*, Brookes Publishing Company, Baltimore, MD.

Macfarlane, A., and S. Luzzadder-Beach (1998), Achieving equity between women and men in the geosciences, *GSA Bull.*, 110(12), 1590–1614.

Mayr, E. (1997), *This Is Biology: The Science of the Living World*, Harvard University Press, Cambridge, MA.

McInnes, K. L., K. J. E. Walsh, G. D. Hubbert, and T. Beer (2003): Impact of sea-level rise and storm surges on a coastal community, *Natural Hazards* 30(2), 187–207.

Mercer, J., I. Kelman, I., K. Lloyd, and S. Suchet-Pearson (2008), Reflections on use of participatory research for disaster risk reduction, *Area*, 40(2), 172–183.

Merton, R. K. (1973), The normative structure of science, in R. K. Merton (Ed.), *The Sociology of Science: Theoretical and Empirical Investigations* (pp. 267–278), University of Chicago Press, Chicago.

Minkler, M., and N. Wallerstein (2010), *Community-based Participatory Research for Health: From Process to Outcomes*, Jossey-Bass, San Francisco, CA.

Moser, S. C., and L. Dilling (2006), *Creating a Climate for Change*, Cambridge University Press, Cambridge, UK.

Moss-Racusina, C. A., J. F. Dovidiob, V. L. Brescollc, M. J. Grahama, and J. Handelsmana (2009), Science faculty's subtle gender biases favor male students, *Proc. Natl. Acad. Sci.*, 109(41), 16474–16479.

National Science Board (NSB) (2012), *Science and Engineering Indicators*, Arlington, VA (NSB 12-01), retrieved November 7, 2013, from http://www.nsf.gov/statistics/seind12.

National Science Foundation (NSF), National Center for Science and Engineering Statistics (2013), *Women, Minorities, and Persons with Disabilities in Science and Engineering: 2013*, Special Report NSF 13-304, Arlington, VA, retrieved November 7, 2013, from http://www.nsf.gov/statistics/wmpd/.

O'Brien, T. L. (2012), Change in academic coauthorship, 1953–2003, *Sci. Tech. Human Val.*, 37(3), 210–234.

Olson, R. (1997), *Don't Be Such a Scientist*, Island Press, Washington, DC.

Oreskes, N., and E. M. Conway (2010), *Merchants of Doubt: How a Handful of Scientists Obscured the Truth on Issues from Tobacco Smoke to Global Warming*, Bloomsbury Press, London.

Pandya, R. E. (2012), A framework for engaging diverse communities in citizen science in the US. *Front. Ecol. Environ.*, 10, 314–317.

Park, P. (1993), What is participatory research? A theoretical and methodological perspective, in Park, P., M. Brydon-Miller, B. Hall, and T. Jackson (Eds.), *Voices of Change: Participatory Research in the United States and Canada* (pp. 1–21), OISE Press, Toronto, Ontario, CA.

Pierotti, R., and D. Wildcat (2000), Traditional ecological knowledge: The third alternative (Commentary), *Ecol. Appl.*, 10(5), 1333–1340.

Polyani, M. (1962), The republic of science: Its political and economic theory, *Minerva* 1, 54–74.

Porticella, N., S. Bonfield, T. DeFalco, A. Fumarolo, C. Garibay, E. Jolly, et al. (2013), *Promising Practices for Community Partnerships: A Call to Support More Inclusive Approaches to Public Participation in Scientific Research*, A Report Commissioned by the Association of Science-Technology Centers.

Poundstone, W. (1999), *Carl Sagan: A Life in the Cosmos*, Henry Holt, New York.

Probst, K., J. Hagmann, M. Fernandez, and J. A. Ashby (2003), *Understanding Participatory Research in the Context of Natural Resource Management: Paradigms, Approaches and Typologies*, Overseas Development Institute (ODI), United Kingdom.

Seymour, E., and N. M. Hewitt (1994), *Talking about Leaving: Factors contributing to High Attrition Rates among Science, Mathematics and Engineering Undergraduate Majors: Final Report to the Alfred P. Sloan Foundation on an Ethnographic Inquiry at Seven Institutions*, Alfred P. Sloan Foundation, University of Colorado, Boulder.

Sarewitz, D., and R. A. Pielke, Jr. (2007), The neglected heart of science policy: Reconciling supply of and demand for science, *Environ. Sci. Pol.*, 10(1), 5–16.

Sills, G. L., N. D. Vroman, R. E. Wahl, and N. T. Schwanz (2008), Overview of New Orleans levee failures: Lessons learned and their impact on national levee design and assessment, *J. Geotech. Geoenviron. Engrg.*, 134(5), 556–565.

Shirk, J., H. Ballard, C. Wilderman, T. Phillips, A. Wiggins, R. Jordan, et al. (2012), Public participation in scientific research: A framework for deliberate design, *Ecol. Soc.*, 17(2), 29.

Stokes, D. (1977), *Pastuer's Quadrant: Basic Science and Technological Innovation*, Brookings Institution Press, Washington, DC.

Strenta, A. C., R. Elliott, R. Adair, M. Matier, and J. Scott (1994), Choosing and leaving science in highly selective institutions, *Res. Higher Ed.*, 35(5), 513–547.

Tervalon, M., and J. Murray-Garcia (1998), Cultural humility versus cultural competence: a critical distinction in defining physician training outcomes in multicultural education, *J. Health Care Poor Underserved*, 9(2), 117–125.

Tobias, S., and M. Fehrs (1991), They're not dumb, they're different: Stalking the second tier, *Physics Today*, 44(81).

Travis, J. (2005), Scientists' fears come true as hurricane floods New Orleans, *Science*, 309, 1656–1659.

Valian, V. (1999), *Why So Slow? The Advancement of Women*, MIT press, Cambridge, MA.

Viswanathan, M., A. Ammerman, E. Eng, G. Garlehner, K. N. Lohr, D. Griffith, et al. (2004), *Community-Based Participatory Research: Assessing the Evidence: Summary*, Evidence Report/Technology Assessment: Number 99, AHRQ Publication Number 04-E022-1, August 2004, Agency for Healthcare Research and Quality, Rockville, MD, retrieved November 7, 2013, from http://www.ahrq.gov/clinic/epcsums/cbprsum.htm.

World Health Organization (WHO) (2003), *Meningococcal Meningitis Fact Sheet No. 141*, retrieved November 7, 2013, from http://www.who.int/mediacentre/factsheets/fs141/en/.

Wisner, B. (2004), *At Risk: Natural Hazards, People's Vulnerability and Disasters*, Psychology Press, East Sussex, United Kingdom.

Yamin, F., A. Rahman, and S. Huq (2005), Vulnerability, adaptation and climate disasters: a conceptual overview, *IDS Bull.*, 36(4), 1–14, Institute of Development Studies, Brighton, United Kingdom.

# 7

# Geoscience Alliance: Building Capacity to Use Science for Sovereignty in Native Communities

Nievita Bueno Watts[1], Wendy Smythe[2], Emily Geraghty Ward[3], Diana Dalbotten[4], Vanessa Green[5], Mervyn Tano[6], and Antony Berthelote[7]

## 7.1. THE GEOSCIENCE ALLIANCE

The Geoscience Alliance (GA) is a national alliance of individuals committed to broadening participation of Native Americans in the geosciences. Its members are faculty and staff from tribal colleges, universities, and research centers; Native elders and community members; industry and corporate representatives; students (K–12, undergraduate, and graduate); formal and informal educators; and other interested individuals.

Mission and Vision of the Geoscience Alliance

We envision a future in which Native Americans are no longer underrepresented in the geosciences. We look to a future where Native scientists take a leadership role in helping to steer our country towards a more sustainable and environmentally ethical relationship with the Earth. To appreciate and advance the geosciences while being respectful of indigenous cultures, we articulate the following values:

• We focus on supporting students, even as we recognize that we all are students.

• We respect both western and indigenous approaches to knowing about the Earth, while recognizing that indigenous approaches historically have been undervalued.

• We believe that there are many paths to being a scientist and many traditions to draw from. Therefore, there is no single best practice; instead, we offer a collection of effective strategies to draw from.

• We will create ways for students to become scientists while holding onto and even strengthening their traditional knowledge.

• We are inclusive: our focus on increasing Native Americans in the geosciences doesn't confine membership to either geoscientists or Native Americans.

• We will explore, make mistakes, forgive and learn together.

The authors of this chapter are all members of the GA and of the GA's "Sustainability" Committee, which is a small group of GA volunteers committed to providing day-to-day leadership of the GA between formal meetings and events. In this chapter, members of the alliance present information pertinent to building capacity in Native communities to use science as a means of securing sovereignty in resource-management decisions on tribal lands. It is our position that this capacity is necessary and can be achieved through the attainment of high school diplomas and higher education degrees in Earth and environmental science by tribal members. In the sections which follow, the authors respond to recommendations of the GA for broadening participation of Native Americans in Geosciences [*Dalbotten*, 2010]: Incorporate traditional knowledge into geoscience education for Native students; understand and address barriers to obtaining undergraduate and graduate degrees in the geosciences; and create culturally appropriate assessment and evaluation in Indian Country. This research is supported by findings from the first Geoscience Alliance National Conference held in Cloquet, Minnesota in September 2010. Since then, members of the GA have been involved in several projects specifically aimed to address the issues and

---

[1] *Director of Academic Programs, Center for Coastal Margin Observation & Prediction, Institute of Environmental Health, Oregon Health & Sciences University, Portland, Oregon*

[2] *K'ah Skaahluwaa, Center for Coastal Margin Observation & Prediction, Institute of Environmental Health, Oregon Health & Sciences University, Portland, Oregon*

[3] *Assistant Professor of Geology, Rocky Mountain College, Billings, Montana*

[4] *Director of Diversity and Broader Impacts, National Center for Earth-Surface Dynamics, St. Anthony Falls Laboratory, University of Minnesota, Minneapolis, Minnesota*

[5] *Director of Higher Education and Diversity, Center for Coastal Margin Observation & Prediction, Institute of Environmental Health, Oregon Health & Sciences University, Portland, Oregon*

[6] *President, International Institute for Indigenous Resource Management, Denver, Colorado*

[7] *Hydrology Program Director, Salish Kootenai College Natural Resources Department, Pablo, Montana*

---

*Future Earth—Advancing Civic Understanding of the Anthropocene, Geophysical Monograph 203*, First Edition.
Edited by Diana Dalbotten, Gillian Roehrig, and Patrick Hamilton.
© 2014 American Geophysical Union. Published 2014 by John Wiley & Sons, Inc.

barriers raised here, and do so with a focus on responding to issues raised in Native American communities under the stress of anthropogenic global changes.

## 7.2. UNDERREPRESENTATION OF NATIVE AMERICANS IN THE GEOSCIENCES

Native Americans are underrepresented in Earth, environmental, geographical, and spatial science and other higher education programs. Compared with other science and engineering fields, the geosciences produce the lowest percentage of minority scientists with bachelor and master's degrees. Underrepresented minorities currently comprise 30.5 percent of the US population, with 15 percent of the population being non-Hispanic Blacks, 14 percent Hispanics, and 1.5 percent American Indian and Alaska Natives. The percentage of geoscience bachelor's degrees conferred on minority students is much less than 30.5 percent. In 2004, for example, Hispanics received 3.3 percent of the bachelor of science degrees, Blacks 1.7 percent, and American Indians 0.8 percent. For master's degrees in geosciences, the percentages drop to 2.3 percent for Hispanics, 1.4 percent for Blacks, and 0.5 percent for American Indians. At the doctoral level in 2004, the rates stayed about the same as for master's degrees, with Hispanics garnering 2.3 percent, Blacks 1.7 percent, and American Indians 0.4 percent [*American Geological Institute (AGI)*, 2009]. This level of underrepresentation and the underlying issues that lead to it are the motivating factors that led to the creation of the GA.

## 7.3. GEOSCIENCE ALLIANCE GOALS

The goals of the GA are to create new collaborations in support of geoscience education for Native American students; establish a new research agenda aimed at closing gaps in our knowledge on barriers and best practices related to Native American participation in the geosciences; increase participation by Native Americans in setting the national research agenda on issues in the geosciences, particularly those that impact Native lands; provide a forum to communicate educational opportunities for Native American students in the geosciences; and to understand and respect indigenous traditional knowledge.

## 7.4. BACKGROUND OF THE GEOSCIENCE ALLIANCE

The GA was formed in 2007 through the efforts of the National Center for Earth-surface Dynamics (NCED). Key partners from Salish Kootenai College, Purdue University, the GLOBE Program, University Consortium for Atmospheric Research (UCAR), and others joined and helped to bring about the first GA meetings, which were held in conjunction with other conferences, such as the American Geophysical Union (AGU), the Society for the Advancement of Chicanos and Native Americans in Science (SACNAS), and the American Indian Science and Engineering Society (AISES) national meetings.

## 7.5. THE GEOSCIENCE ALLIANCE NATIONAL CONFERENCES

In 2010, the National Science Foundation sponsored the first GA conference through an "Opportunities for Enhancing Diversity in the Geosciences (OEDG)" Award (NSF GEO 0939753). This conference brought together more than 100 individuals from tribal and other institutions—with more than half students—to discuss barriers to broadening participation and ways to overcome them. In addition, students had the opportunity to meet with program directors and hear about research opportunities in the geosciences. This led to many productive outcomes, including a summary report on the conference discussions [*Dalbotten*, 2010]; students participating in research programs, new research, and education collaborations; and a dissertation published on the topic of broadening participation of Native Americans in the geosciences [*Bueno Watts*, 2011]. The conference report and dissertation are important for being the first studies to address underrepresentation of Native Americans specifically in the geosciences and to examine key barriers and potential solutions to this problem.

The second GA national conference was held in March 2012 with the topic of *Home Places, Local Landscapes, Traditional Knowledge, and Modern Technologies*, with support from the National Science Foundation and NASA. It was located at Salish Kootenai College (SKC), a tribal college in Pablo, Montana. Several sophisticated research techniques and data management and visualization tools were highlighted. Keeping in mind the circle of learning principles that guide the GA (i.e., everyone teaches and everyone learns), these workshops were structured as a dialogue with positive feedback loops; participants not only explored the technologies and their potential for use in Native communities but were also engaged in helping the institutions who provided these tools to better understand and meet the needs of Native communities. Climate change and its impacts on Native reservations was a crosscutting concern of the conference and appeared as a theme in several of the posters and presentations of the students. Clearly, Native American students are concerned about the effects of anthropogenic change on their homelands and cultural resources. This chapter discusses research presented at the 2012

GA conference that was motivated by discussions at the 2010 GA conference as well as highlights the progress that has been made through collaborations born of the alliance. These studies are discussed in detail.

## 7.6. INCORPORATING TRADITIONAL KNOWLEDGE IN GEOSCIENCE EDUCATION

Native American and Alaska Native K–12 students of all ages face numerous challenges both socially and academically. In 2003, the US Commission on Civil Rights found that Native students commonly experienced "deteriorating school facilities, underpaid teachers, weak curricula, discriminatory treatment, and outdated learning tools" [*Berry et al.*, 2003]. There is a push for revitalization occurring in Native education and for traditional knowledge to be taught alongside other current science concepts to provide culturally relevant curriculum for STEM disciplines [*Brave Heart and DeBruyn*, 1998; *Brave Heart*, 1999]. Indian Country is poised to nurture a new generation of Native scientists and natural resource managers who can guide their communities to a sustainable future in the face of anthropogenic change.

Traditional knowledge, also referred to as indigenous knowledge or traditional environmental/ecological knowledge, refers to longstanding traditions and practices, encompassing the knowledge, culture, and spirituality of Native Americans, Alaska Natives, Native Hawai'ians, and other indigenous peoples around the world. Historically traditional knowledge has been orally passed on for generations and expressed through stories, legends, rituals, and songs [*Berkes*, 1993]. Traditional knowledge is founded on empirical observations, applied practices, and a lifestyle in which competency in natural resource science and management is essential for and defined by survival [*Simpson*, 2002; *Barnhardt and Kawagley*, 2005]. Indigenous cultures have collected and maintained extraordinary amounts of comprehensive knowledge through many generations of living in a specific region and through their spiritual ties to the environment. Coupling traditional knowledge systems with other science concepts provides a greater depth of understanding as to how natural ecosystems are changing and can augment resource management practices as managers respond to global climate change.

Traditional knowledge systems are holistic and interweave science, ethics, religion, philosophy, medicine, psychology, and economics as all part of the same body of knowledge [*Barnhardt and Kawagley*, 2005]. This holistic approach used by traditional knowledge provides Native students with a foundation from which to understand scientific concepts such as geologic time, climate change, ecology, and evolution.

### 7.6.1. Obstacles in Geoscience Education for Native Americans at the K–12 Level and the Role Traditional Knowledge Can Play in Removing These Obstacles

The difficulty of school systems to effectively educate Native students is a result in large part of four systemic, interrelated obstacles resulting in a divide that Native students and STEM professionals find challenging to overcome. The first is an attitude that traditional knowledge holds little or no value [*Haig-Brown*, 1995; *Deyhle and Swisher*, 1997; *Kawagley and Barnhardt*, 1999a; *Deyhle*, 2010]. This attitude manifests itself when culturally disengaged teachers attempt to challenge ethnoscientific knowledge, leaving Native students feeling pressured to choose between the correctness of one knowledge system over the other. Naturally many students choose the knowledge system with which they are familiar and as a result many withdraw from active participation in school, whereas other students leave school altogether [*Deyhle and Swisher*, 1997; *Barnhardt and Kawagley*, 2004; *Deyhle*, 2010].

The second obstacle is that non-native teachers are typically viewed as outsiders, and Native communities and students may have an us-versus-them relationship with teachers and administrators. This mutually exclusive type of relationship is detrimental to both the school staff and more importantly the students. Rather than continue this negative cycle of exclusion, solutions need to be found that overcome this obstacle for the future of Native students. One solution would be for the tribal community to embrace educators and school administrators, simultaneously implementing a "who we are," "what we stand for," and "why it is important" framework to assist non-local educators in understanding community and culture. The divide between teachers and Native students can be bridged through the combined commitment of mentoring by culturally competent tribal members and the acceptance of teachers and school administrators into Native communities. Educators who are welcomed are better suited to interact positively with the students, parents, community, and tribal members. Without awareness of the local culture, it is virtually impossible for culturally relevant curriculum to be taught. Therefore, it is imperative that Native communities work with educators and administrators, both Native and non-native, to develop culturally relevant STEM curriculum to ensure the educational success of their students.

The third obstacle is teaching strategies that contradict the learning style of many Native students, thereby setting up a psychological power struggle between traditional ways of teaching and learning and western educational system requirements [*Nelson-Barber and Estrin*, 1995; *Cajete*, 1999; *Kawagley and Barnhardt*, 1999b; *Barnhardt and Kawagley*, 2005]. Historically, Native peoples have not had the power to define what constitutes education, much less science education [*Lomawaima*, 2000]. Pedagogies commonly used in geoscience courses expect students to adopt and understand an unfamiliar and abstract approach, which is in direct opposition to the situational, concrete knowledge emphasized in Native cultures. Although Native cultures vary, *Cajete* [1999] describes some common attributes of Native learners, including quietness and patience, a cooperative social orientation, a holistic and spiritualistic view of nature, and a nonlinear sense of time. Native students tend to prefer concrete, contextualized lessons in which they observe before practicing. Pedagogical approaches can accomplish this through the use of STEM curriculum that couples traditional knowledge and language with current science curriculums that include both a field and laboratory component.

The fourth obstacle resides in the inaptitude of Native people to identify as scientist, teacher, or any other position that requires higher education, hierarchical, or noncommunal thinking. This inaptitude is a result in large part of the historical trauma inflicted on Native peoples by governments that sought to erase cultural identity [*Lomawaima*, 2000]. Reforms in the 1970s and thereafter replaced federal and state schools, which focused on assimilation of Native peoples into western society [*Dlugokinski and Kramer*, 1974; *Lomawaima*, 1995; *Deyhle and Swisher*, 1997], with locally controlled schools, schoolteachers, and administrators.

Tribal elders experienced this cultural assault firsthand and survived through self-determination and courage; however, successive generations have also suffered as a result of the long-term effects of this cultural assault. It is imperative that we not only educate Native youth using traditional knowledge but also demonstrate that Native people can be the scientists, engineers, teachers, and educated tribal members within their own communities. Native communities have only recently begun to revitalize through healing [*Brave Heart and DeBruyn*, 1998; *Brave Heart*, 1999] by reclaiming traditions, such as canoeing, hunting, and languages that have not been practiced for generations [*Cajete*, 1999; *Deyhle and Swisher*, 1997; *Simpson*, 2002; *Deyhle*, 2010]. Science education programs can use the momentum of this movement by adopting culturally relevant and student-centered pedagogies that recognize the current state of Native cultures, both accommodating the ways in which Native peoples have traditionally learned and recognizing how Native cultures have evolved since the 1900s [*Cajete*, 1999].

Overcoming obstacles in Native education is no easy task. However, these obstacles can be overcome through recognizing the value in traditional knowledge and incorporating it into novel pedagogical approaches. Such approaches are student centered; inquiry based; respect local traditions, knowledge, and culture; and meet community-identified needs [*Bueno Watts and Smythe*, 2013; *Hugo et al.*, 2013].

## 7.7. REMOVING BARRIERS TO BROADENING PARTICIPATION IN UNDERGRADUATE AND GRADUATE EDUCATION

Herein, we summarize the results from Bueno Watts's dissertation research on "Broadening Participation of Native Americans in the Earth Sciences," which was both inspired by her participation in the GA and facilitated by interviews with several GA members. The dissertation reports the results of semi-structured interviews, which ranged from 30 to 90 minutes in length, designed to discover what factors were barriers to attaining a degree, and what factors helped with completion of a geoscience program of study. These interviews were conducted with fifteen Native Americans who had already achieved, or were in the process of attaining, a postsecondary degree in Earth or environmental science and 10 directors of programs designed to increase the number of Native Americans graduating from Earth and environmental science degree programs. The interviews were analyzed qualitatively following methods outlined in *Miles and Huberman* [1984].

### 7.7.1. Demographics of Study Sample

Twelve of the 15 Native participants interviewed (80 percent) were female and three (20 percent) male. Eleven (73 percent) of the participants were first-generation college students, and the rest (27 percent) reported that one or both of their parents had been the first in the family to attend college, making them second-generation college students. About half (53 percent) were nontraditional students, defined here as someone who found themselves in school 20 years after high school graduation. Each nontraditional student either took an extended period of time off from school to work between degrees (75 percent) or worked full-time while pursuing a degree (25 percent), thereby extending the time required for completion.

Seven out of 15 participants (47 percent) started their college education at a public four-year institution; an additional 3 (20 percent) began at a private, religious, four-year institution. Five out of 15 (33 percent) completed their bachelor's degree without transferring or taking an extended break. Three out of the 15 (20 percent) participants began college at a local community college, and 2 out of 15 (13 percent) began their studies at a tribal college. Seven out of 15 (47 percent) of participants did not complete their initial choice of programs. Only 3 of the 15 (20 percent) participants started their educational path in a geoscience or environmental Earth science field.

Combined, the 15 participants had completed 6 associate degrees, 12 bachelor degrees (3 participants were undergraduate students), 7 master's degrees (with 2 enrolled in master's programs), and 2 doctorate degrees (with 4 more enrolled in doctoral programs).

### 7.7.2. Barriers to Completing a Geoscience Degree

The interviews uncovered a number of barriers that could be grouped into general themes: barriers to completing a college degree, impediments to making a decision to study Earth Science, and factors that impede progress in Earth science programs.

Many of these barriers are certainly not unique experiences of Native American students, but as expressed, compounded into nearly unbearable burdens for many students. Several factors were barriers to completing a college degree. For example, without financial aid resources 80 percent of the participants would not have been able to afford a college degree. In addition, pressures from familial obligations often interrupted studies. Some familial obligations, such as needing to go home for funerals or other ceremonial obligations, posed problems for some members of this group. In other cases students had families that they needed to rush home for, precluding them from taking advantage of programs and opportunities incompatible with the life styles of students who are also working parents.

Many participants in the study also described experiencing debilitating physical and mental health issues. Some mental health issues that arose, like those associated with feelings that a student "wasn't good enough" to be a graduate student in the Earth science department, or the depression and feelings of helplessness that occurred when work piled up too high, are not strictly conditions of being Native American. They were reported by a third of the women interviewed and were exacerbated by departmental environments that were particularly hostile to women. A good advisor or other support person, however, helped these students deal with both physical and mental health issues.

A second set of factors were impediments to making a decision to study Earth science. The most commonly reported was unfamiliarity with geoscience as a field of study or career path—80 percent of the participants talked about geoscience not being known in tribal communities as either a field of study or a career choice. Even if geoscience was selected as a field of study, however, the curriculum offered was often seen as being irrelevant to the practical needs of the community, and course names unintelligible to non-geologists, and the degree program was therefore discounted as a way of solving the problems found in Native homelands. The importance of this barrier cannot be overemphasized because an overwhelming percentage (93 percent) of participants expressed a desire to work on environmental geology issues. Particularly for re-entry students who were non-traditional in age, desire to heal the land was a motivating factor in their decision to return to school. Unfortunately, even after enrolling in a geoscience program, students found that the career they thought they were preparing for, where they would be able to use the skills learned to solve environmental problems back home, did not line up with the focus or program of the department, and this was a major disappointment. Locations of institutions themselves often pose an additional barrier because many are inaccessible geographically, and large segments of the Native population do not have access to Earth science courses near their homes.

A third set of factors impeded academic progress through Earth science academic programs. The most often expressed academic barrier reported was that of inadequate preparation in mathematics (73 percent). But it was not because Native students could not "do" math—most eventually took Calculus I and Calculus II and passed—but the quality and availability of mathematics instruction was the institutional barrier in this case, at *all* levels of the educational system. The biggest barrier this problem poses is the spiraling effect that bad grades in calculus and chemistry (another frequently mentioned barrier course), have on a student's grade point average, subjecting them to academic probation and making them ineligible for funding (another institutional barrier).

Lack of academic information and counseling was also a barrier to progress for students. Nontraditional students reported that the failure of both Earth science departments and university counseling systems to provide appropriate guidance for older students was problematic. They reported having to seek out information on graduate school, programs, and other opportunities and were not approached by faculty with opportunities for internships or scholarships as often as their younger peers. Some even felt that student counseling programs in place at the university level were not designed to accommodate their needs. Having a departmental mentor to talk to became imperative for these students' success.

Intradepartmental relationships were another barrier to success within the day-to-day workings of departmental politics. Conflicts between subfields were especially detrimental to student success when the advisor of a student was in a new, nontraditional field that the other members of the department did not support (i.e., biogeochemistry).

None of the participants described feeling discrimination in Earth science departments by virtue of the fact that they were Native, yet many of the women reported harassment by faculty members. Harassment was sometimes so intense that it affected the student's relationships with other students and jeopardized their ability to complete their program of study. The situation became so intolerable in at least two cases that the student moved either to another department or another university. It must be understood that many Native women are respected leaders within their community and have grown up in the outdoors, often performing the same tasks that men do in white rural societies. Some of the women interviewed expressed surprise to find themselves in a field that was considered to be non-traditional for women because they came into the program assuming they were capable of doing anything a man could. Harassment was more reported by nontraditional students than by those who were of traditional age, and several incidents occurred during the last decade (2000s), indicating that this inappropriate behavior is not a thing of the past.

Some students, having entered geoscience programs, reported that the prevalence of western scientific perspective to the exclusion of all others became cognitively problematic. There was no acknowledgement of the local peoples or their knowledge of the land, even in schools with a high concentration of Native students. Furthermore, 11 out of the 15 participants (73 percent) described themselves as thinking holistically when approaching problems, which is contrary to methodologies that tend to break knowledge into smaller pieces for examination.

One promising move toward increasing geoscience degrees earned by Native Americans is the introduction of two- and four-year geoscience degree programs at tribal colleges and universities. In this way, many of the barriers described previously are removed. SKC recently announced the first two- and four-year geoscience degree offered at any tribal college or university in the country, offering new opportunities for Native students to participate in the geosciences.

### 7.7.3. The Salish Kootenai College Hydrology Degree Program

SKC received approval in the fall of 2010 from the Northwest Commission on Colleges and Universities to offer both associate and bachelor of science in hydrology, the first such degree programs among the tribal colleges and universities in North America. The hydrology program is aligned with the SKC strategic plan goal to "become a center of science education with an emphasis on Native American worldview and application of science to indigenous issues." SKC's hydrology program offers interdisciplinary study of physical, chemical, and biological water resources and their management. These efforts will significantly increase the number of Native Americans receiving degrees in the geosciences.

Water is a key natural resource in today's world and is a resource that is increasingly coming under stress as a result of climate change. Climate factors are also expected to play a growing role in increasing vulnerability to natural hazards such as floods or hurricanes. Tribal resource managers with skills in hydrology are needed to assess vulnerability of tribal lands to climate change and develop a plan for management. The summer 2012 flooding, which took place across the Fond du Lac Band of Lake Superior Chippewa reservation in Northeastern Minnesota, is an example of the devastating effect natural hazards can have on a reservation, impacting roads, fishing, forestry, wild ricing, housing, and health. Thomas Howes, Natural Resources Manager for Fond du Lac reservation noted: "we are gravely concerned about the future of *manoomin* (wild rice) in the face of climate change. Given the sensitive nature of *manoomin* to hydrologic events and the drastic landscape changes in the region that *manoomin* grows, we need additional hydrologists and researchers to explore wetland and ditching alteration remedies" [*personal communication*, May 9, 2013].

There is an urgent need for improved teaching and learning at tribal colleges and universities related to geoscience issues to support the next generation of future Earth land and water resource managers for Native American communities and across the United States. Twenty-three of the 37 tribal colleges and universities offer degrees, certificates, or courses in natural resources or environmental sciences, but these programs, which largely result in certificates of completion or associates degrees after completion, are in a constant state of flux. Most do not lead to transfers into four-year geoscience programs and therefore are not translating into increased numbers of bachelor's degrees in the geosciences.

Nationally, according to the 2008–2009 Occupational Outlook Handbook of the Bureau of Labor Statistics, the employment forecast for US hydrologists will experience a 18 percent increase from 2010 to 2020. An evaluation of the geoscience workforce indicates that there is a growing deficiency between the developing and needed workforce [*AGI*, 2009]. The nation will need new appraisals of water availability in the next decade that link both water quality and quantity; track changing flow, use, and storage of water, as well as models and predictive tools to guide its management decisions [*USGS*, 2007]. This has important implications for Native American reservations in the face of increasing anthropogenic changes.

The SKC hydrology program has the potential to have impacts nationwide as students take up jobs within the state or return to their reservations. Developing broader use of technology in the hydrology curriculum at SKC will prepare Native American students to face the challenges found in modern water resources research and management. Native American students will be better prepared to effectively participate in any discipline or profession where advanced geospatial, water quality, and water quantification tools have become an important and necessary job qualification. Graduates equipped with the skills to use current technology in research and management will greatly aid tribal and nontribal agencies for which they will work.

The long-term impact of SKC's hydrology programs will culminate in Native hydrologists who can provide the unique perspective of Native peoples on natural resources while taking advantage of the most current technologies. Particularly, students will continue into professions or graduate education with knowledge of water-related issues including water rights, agriculture, environmental hydrohealth, beliefs, and spirituality related to water, and sustainability of water resources while having the advantage of developed proficiencies in current technologies. The purpose of including Native perspectives goes beyond a simple appreciation of Native culture and beliefs; SKC objectives are both to empower Native communities through a sharing of knowledge and experience and to enhance cross-cultural understanding and respect for different approaches to water and water development. Both cultural perspectives and technological objectives have practical and tangible expressions in the realm of improved legal frameworks, agricultural practices, water quality, health, public relations, water rights, sovereignty and other global, international, and local management concerns. The SKC hydrology degree programs strives to engage Native peoples in expressing unique perspectives on water priorities and on what constitutes "improvements" and "progress" for their societies while taking advantage of technological offerings and real-time data sets.

## 7.8. CULTURALLY APPROPRIATE, NATIVE-FOCUSED ASSESSMENT AND EVALUATION FOR THE GEOSCIENCES

A clear call from the first GA conference was for culturally sensitive or "native-friendly" practices that work to incorporate Native cultural perspectives and practices when planning assessment or evaluation of programs. Researchers have expressed concern about the cultural validity of science assessments in particular, arguing that sociocultural context influences both the interpretation and solution of questions [*Solano-Flores and Nelson-Barber*, 2001; *Lee and Luykx*, 2007]. *Solano-Flores and Nelson-Barber* [2001, p. 555] define cultural validity as "the effectiveness with which science assessment addresses the sociocultural influences that shape student thinking and the ways in which students make sense of science items and respond to them." The sociocultural context for Native communities is place, particularly for the study of natural sciences. Place is defined here as both the sociocultural landscape shaped by people and the physical landscape formed by Earth's processes [*Relph*, 1976; *Tuan*, 1977; *Gould and White*, 1986; *Cajete*, 2000; *Deloria and Wildcat*, 2001; *Aikenhead and Michell*, 2011].

Researchers have proposed infusing context to improve the effectiveness of assessment and evaluation practices [*Solano-Flores and Nelson-Barber*, 2001; *Nichols and LaFrance*, 2006; *LaFrance and Nichols*, 2010]. This type of assessment and evaluation design and implementation becomes an inherently collaborative process. Both assessment and evaluation methodologies resulting from this process are community focused and dependent on the program goals and cultural values of the stakeholders involved. An exemplar for contextualizing assessment is presented here and discussed in terms of its applications to evaluation practice.

The NSF-funded Cultural Validity in Geoscience Assessment project collaborative (NSF GEO-1034909 and GEO-1034926), in an effort to design place-based, culturally informed science assessment, identified experts from the Blackfeet and Navajo (Diné) communities who were involved in science education or had expertise in Native culture and language. The selected experts were involved in the entire process of assessment design. The expert group, along with tribal college faculty and students, were surveyed to ask them what geoscience topics were important to their community to identify important and culturally relevant geoscience concepts to be used in assessment. Their responses focused the assessment on relevant, place-specific content. Topics such as water, glaciers, and mountains were particularly relevant to Blackfeet community participants. Some topics identified embodied an Earth system science perspective, highlighting the interconnectedness of Earth materials and processes [*Ward, et al.*, 2011; *Ward, et al.*, 2014].

The Blackfeet expert group convened to discuss the survey results and to author a suite of open-ended assessment questions that would incorporate the place-specific content identified from the surveys. The questions focused on the elements of Earth, wind, water, and fire and were aligned with relevant Earth science literacy principles [*Earth Science Literacy Initiative (ESLI)*, 2009] as well as the Blackfeet Education Standards developed by the tribal college [*Blackfeet Community College Rural Systemic Initiative*, 2005]. Four questions were selected for piloting with science students attending Blackfeet Community College to collect their responses. The responses to the open-ended questions provided data for the expert group to develop closed-response assessment questions. The expert group identified themes using content analysis of the student responses and grounded the new assessment questions in the language and content provided by students for the open-ended assessment questions.

The product was a suite of geoscience assessment questions aligned with Earth science literacy principles and Blackfeet education standards that address important and culturally relevant geoscience content for Blackfeet students. The newly constructed assessment suite provides a culturally validated measure of conceptual understanding for students from this community as a result of the collaborative process. The assessment process outlined previously aligns with the evaluation framework designed by American Indian Higher Education Consortium [*LaFrance and Nichols*, 2009] in that it is collaborative, contextualized, and in alignment with the cultural values of the project stakeholders. Hence, these methods of assessment can be adapted for the evaluation of new and ongoing research and education projects on the reservation.

Identifying meaningful outcomes for projects comes easily for these cultural experts. These outcomes often encompass a broad range of variables from gathering information from cognitive, behavioral, and affective domains, to increasing access to STEM experiences and giving back to the community. External evaluation can select data gathering methods that align with community preferences when partnerships are formed with local evaluators [*LaFrance and Nichols*, 2009; *Kirkhart et al.*, 2011]. For example, if the local evaluator indicates that participants prefer oral administration of a survey or to have their answers audio-recorded rather than handwritten, data collection methods are able to accommodate these preferences, facilitating participant feedback.

Another effective practice is to share the data interpretation with participants to ensure validity, to retain the original voice, and to gather any additional feedback before reporting project findings [*LaFrance and Nichols*, 2009]. Assessment and evaluation practices in Native communities require a mixed methods approach. Practices that are collaborative, community driven, and contextualized offer rich and culturally valid approach to assessment and evaluation.*

### 7.9. CONTINUING ISSUES

Native Americans continue to be underrepresented in the geosciences, despite ongoing efforts to provide pathways into geoscience careers. These students have a crucial role to play in the future management of our nation's land and water resources, both on the reservation and across the country. Vast resources have been poured into creating technologies that help us understand and respond to the challenges that face future Earth, but these technologies are underused, particularly by Native communities. Despite recent investments, there continues to be a technology gap in reservation communities. And even in cases where computers and Internet are present, there is still an information and applications gap—the potential for these sophisticated tools to solve local problems is not realized by the potential user community.

### 7.10. FUTURE OF THE GEOSCIENCE ALLIANCE

The GA has worked between conferences and meetings to bring in new members and create multiple avenues for continued participation in the alliance. We have a blog, a Facebook group, and a listserv, all focused on providing information about opportunities and news related to the geosciences. Many organizations, including NSF Research Experience for Undergraduates programs, the American Geophysical Union, federal groups such as NASA, NOAA, and the USGS, and others send information to the GA for distribution to our members. Members are also active on social networks, sharing stories of issues related to indigenous groups around the globe. Our goal for the near future is to continue to build our membership while we develop a plan to support sustainability of the GA in the future. We hope you will join us in our endeavors.

### REFERENCES

Aikenhead, G., and H. Michell (2011), *Bridging Cultures: Scientific and Indigenous Ways of Knowing Nature*, Pearson Canada, Don Mills, Ontario.

American Geological Institute (AGI) (2009), Status of the Geoscience Workforce. American Geological Institute: Alexandria, VA, retrieved from http://www.agiweb.org/workforce/.

Barnhardt, R., and A. O. Kawagley, (2004), Culture, chaos, and complexity: Catalysts for change in indigenous education, *Cult. Survival Quart.*, 27(4), 59.

Barnhardt, R., and A. O. Kawagley (2005), Indigenous knowledge systems and Alaska Native ways of knowing, *Anthro. Ed. Quart.*, 36(1), 8–23.

Berkes, F. (1993), Traditional ecological knowledge: Concepts and cases, in Julian T. Inglis (Ed.), *Traditional Ecological Knowledge in Perspective*; International Program on Traditional Ecological Knowledge and International Development Research Center, Ottawa, Canada.

Berry, M. F., C. Reynoso, J. C. Braceras, C. Edley, P. N. Kirsanow, E. M. Meeks, et al. (2003), *A Quiet Crisis: Federal Funding and Unmet Needs in Indian Country*, U.S. Commission on Civil Rights, Washington, DC.

---

*Co-author Ward would like to thank the numerous participants from Diné College, Blackfeet Community College, and Arizona State University; and members of the Blackfeet and Navajo communities for their work on the assessment project. This project is supported by the National Science Foundation (NSF GEO-1034909). Any opinions, findings, and conclusions or recommendations expressed in this material are those of the authors and do not necessarily reflect the views of the National Science Foundation.

Blackfeet Community College Rural Systemic Initiative (BCCRSI) (2005), *Blackfeet Education Standards Implementation Guide*: Blackfeet Community College, Browning, MT.

Brave Heart, M. Y. H. (1999), Oyate Ptayela: Rebuilding the Lakota nation through addressing historical trauma among Lakota parents, *J. Human Behav. Soc. Environ.*, 2(1-2), 109–126.

Brave Heart, M. Y. H., and L. M. DeBruyn (1998), The American Indian Holocaust: Healing historical unresolved grief, *Amer. Indian Alaska Native Ment. Hlth. Res.*, 3(2), 7–26.

Bueno Watts, N. (2011), Broadening the Participation of Native Americans in Earth Science, (Doctoral dissertation), retrieved from Pro-Quest, UMI Number: 3466860.

Bueno Watts, N., and W. F. Smythe (2013), It takes a community to raise a scientist: A case for community-inspired research and science education in an Alaska Native community, *Currents*, 29 (2).

Cajete, G. A. (1999), The Native American learner and bicultural science education, in K. C. Swisher and J. W. Tippiconnic, III (Eds), *Next Steps: Research and Practice to Advance Indian Education*, (pp. 135–160), Eric Clearinghouse on Rural Education, Denver, CO.

Cajete, G. (2000), *Native Science: Natural Laws of Interdependence*, Clear Light Publishers, Santa Fe, NM.

Dalbotten, D. (2010), *OEDG Planning Grant: Alliance for Broadening Participation of Native Americans in the Geosciences Conference.* Retrieved from http://geosciencealliance.wordpress.com/conferences/2009-conference/.

Deloria, V., Jr., and D. Wildcat (2001), *Power and Place: Indian Education in America*, Fulcrum Resources, Golden, CO.

Deyhle, D. (2010), Navajo youth and Anglo racism: Cultural integrity and resistance, *Harvard Educ. Rev.*, 65(3), 403–445.

Deyhle, D., and K. Swisher (1997), Research in American Indian and Alaska Native education: From assimilation to self-determination, *Rev. Res. Ed.*, 22, 113–194.

Dlugokinski, E., and L. Kramer (1974), A system of neglect: Indian boarding school, *Am. J. Psych.*, 131, 670–673.

Earth Science Literacy Initiative (ESLI) (2009), Earth Science Literacy Principles: The Big Ideas and Supporting Concepts of Earth Science, retrieved November 7, 2013, from http://earthscienceliteracy.org/es_literacy_22may09.pdf.

Gould, P., and R. White (1986), *Mental Maps* (2nd Ed.), Allen and Unwin, Boston.

Haig-Brown, C. (1995), "Two worlds together": Contradiction and curriculum in First Nations adult science education, *Anthro. Ed. Quart.*, 26(2), 193–212.

Hugo, R. C., W. F. Smythe, S. M. McAllister, B. Young, B. Maring, and A. Baptista (2013), Lessons learned from a geoscience education program in an Alaska Native community. *J. Sustain. Educ.*, 5(2151–7452).

Kawagley, A. O., and R. Barnhardt (1999a), Alaska Native education: History and adaptation in the new millennium, *J. Am. Ind. Ed.*, 39(1), 31–51.

Kawagley, A. O., and R. Barnhardt (1999b), Education indigenous to place: Western science meets native reality, in G. A. Smith (Ed.), *Ecological Education in Action*, (pp. 117–140) State University of New York Press, New York.

Kirkhart, K., J. LaFrance, and R. Nichols (2011), Improving Indian Education through indigenous evaluation: Paper presented at the annual meeting of the American Educational Research Association, New Orleans, LA, April 8–12.

LaFrance, J., and R. Nichols (2009), *AIHEC Indigenous Evaluation Framework*, American Indian Higher Education Consortium, Alexandria, VA.

LaFrance, J., and R. Nichols (2010), Reframing evaluation: Defining an indigenous evaluation framework, *Can. J. Prog. Eval.*, 23(2), 13–31.

Lee, O., and A. Luykx (2007), Science education and student diversity: Race/ethnicity, language, culture, and socioeconomic status, in S. K. Abell and N. G. Lederman (Eds.), *Handbook of Research in Science Education* (2nd ed., pp. 171–197), Lawrence Erlbaum Associates, Mahwah, NJ.

Lomawaima, K. T. (1995), *They Called It Prairie Light: The Story of Chilocco Indian School.* University of Nebraska Press, Omaha.

Lomawaima, K. T. (2000), Tribal sovereigns: Reframing research in American Indian education, *Harvard Ed. Rev.*, 70(1), 1–21.

Miles, M. B., and A. M. Huberman (1984), *Qualitative Data Analysis: A Sourcebook of New Methods*, Sage Publishing, Thousand Oaks, CA.

Nelson-Barber, S., and E. T. Estrin (1995), Bringing Native American perspectives to mathematics and science teaching, *Theor. Pract.*, 34(3), 174–185.

Nichols, R., and J. LaFrance (2006). Indigenous evaluation: Respecting and empowering indigenous knowledge, *Tribal Coll. J.*, 18(2), 32–35.

Relph, E. (1976), *Place and Placelessness*, Pion Limited, London.

Simpson, L. (2002), Indigenous environmental education for cultural survival, *Canadian J. Environ. Ed.*, 7(1), 13–26.

Solano-Flores, G., and S. Nelson-Barber (2001), On the cultural validity of science assessments, *J. Res. Sci. Teach.*, 38(5): 553–573.

Tuan, Y-F. (1977), *Space and place: The perspective of experience*, University of Minnesota Press, Minneapolis.

U.S. Geological Survey (2007), *Facing tomorrow's challenges—U.S. Geological Survey science in the decade 2007–2017*, U.S. Geological Survey Circular 1309, USGS, Reston, VA.

Ward, E. M. G., S. Semken, and J. C. Libarkin (2011), Collaborative development of place-based, culturally informed geoscience assessment, *Geol. Soc. Amer. Abstr. Progs.*, 43(5), 75.

Ward, E. M. G., S. Semken, and J. C. Libarkin (2014), The design of place-based, culturally informed geoscience assessment, *Journal of Geoscience Education*.

# 8

## New Voices: The Role of Undergraduate Geoscience Research in Supporting Alternative Perspectives on the Anthropocene

### Diana Dalbotten[1], Rebecca Haacker-Santos[2], and Suzanne Zurn-Birkhimer[3]

Involving undergraduate students in research has long been considered one of the best ways to inspire them to continue with careers in STEM [*Crowe and Brakke*, 2008; *Hancock and Russell*, 2008; *Brownell and Swaner*, 2010; *Thurgood et al.*, 2010]. Now as we face one of our largest challenges ever—dealing with the known and as yet unforeseen impacts of the Anthropocene on future quality of life—we desperately need to inspire the next generation of scientists to not only participate in STEM careers, particularly those related to the geosciences, but to work on the understanding of climate change impacts and adaptation. Recent reports by the National Research Council (NRC) highlight the importance education plays in preparing for the future in an anthropocenic era [*NRC*, 2010, 2012a, b]. In a 1996 report, the National Science Foundation's (NSF) Geoscience Directorate identified the value of Research Experiences for Undergraduates (REUs) for promoting learning in the geosciences:

> Fieldwork is a defining aspect of the geosciences. It therefore needs to assume a central role in undergraduate geoscience education. One of the best vehicles for providing undergraduate students with field opportunities is the Research Experiences for Undergraduates (REU) program. Anecdotal evidence indicates that the REU program has been notably successful in providing hands-on experience and cooperative learning—both in the field and also in the laboratory. It has attracted numerous young people to the geosciences, including many from underrepresented groups, and has also proven to be an important cornerstone for many others who have gone into other professional fields [*NSF*, 1996, p. 33].

At some universities, research experiences have always been a part of the undergraduate experience, both informally and in formal programs. Historically, students have participated in these programs at their home institutions where a single student is placed into the laboratory of a faculty member. They relied on a strong university research program, faculty with time to oversee the student's research, and a student with the willingness to seek out, or access to, the opportunity. In 1958, NSF began to foster these experiences through the Undergraduate Research Program (URP). This program allowed universities to competitively apply for funding to host teams of students in a summer research program. Research that followed confirmed the value of these experiences for promoting students to participate in STEM and to move on to graduate programs. In 1978 the Council for Undergraduate Research (CUR) was formed to promote research opportunities as an integral part of the undergraduate experience [*CUR*, 2009].

The URP was suspended in 1978 for review, and then replaced by the REU program in 1987 [*Schowen*, 2002]. Today, the REU program and other research programs sponsored through organizations such as National Institutes for Health (NIH), National Oceanic and Atmospheric Administration (NOAA), NASA, and US Geological Survey (USGS) have created myriad summer research programs for hundreds of undergraduate students every year. Although the mentoring skills and

---

[1] Director of Diversity and Broader Impacts, National Center for Earth-Surface Dynamics, St. Anthony Falls Laboratory, University of Minnesota, Minneapolis, Minnesota
[2] SOARS Program Director, Head of Undergraduate Education, UCAR Science Education, University Corporation for Atmospheric Research, Boulder, Colorado
[3] Associate Professor, Saint Joseph's College, Department of Mathematics, Rensselaer, Indiana

---

*Future Earth—Advancing Civic Understanding of the Anthropocene, Geophysical Monograph 203*, First Edition.
Edited by Diana Dalbotten, Gillian Roehrig, and Patrick Hamilton.
© 2014 American Geophysical Union. Published 2014 by John Wiley & Sons, Inc.

enthusiasm of the research advisor are still a key element of the quality of the experience, a variety of supports have been added to programs, including writing support, library tutorials, ethics programs, Graduate Record Examination (GRE) prep courses, and others. Participation in research as an undergraduate is now viewed as an integral part of the undergraduate experience by university faculty, administrators, and future employers, especially in STEM fields, and has become an important consideration for entry to a graduate program in STEM. In addition, young faculty who were themselves participants in undergraduate research programs are now eager to share this opportunity with future generations.

The challenge moving forward is to create undergraduate research programs that maintain the benefits of the past—giving students the opportunity to participate in hands-on cutting edge research projects, fostering movement of the strongest STEM students into graduate programs, and making sensible use of faculty time in the process, while expanding the programs' reach and including students from backgrounds that traditionally have not participated in geoscience research. As we are facing the challenges of the Anthropocene, we need a scientific workforce representative of the US population, especially including members of groups who are already and will be disproportionally impacted by current and future environmental change. Studies have shown that students from traditionally underrepresented groups shy away from the geosciences because they do not see it as a field that has a direct positive impact on their communities [*Seymour and Hewitt*, 1997; *Bembry et al.*, 2000]. The Anthropocene might change this. Many underserved communities already disproportionally feel the impacts of a changing environment, such as poorer urban areas susceptible to heat waves, poorer semi-rural areas frequently devastated by tornados, or American Indian lands suffering from drought. Our science needs students from these communities to participate and at the highest level, in research and academia. As participation in REUs has become an unwritten prerequisite for graduate school admission, it is important to make sure that programs are inclusive, providing opportunities to those traditionally underrepresented in the sciences and reaching out to students at two- and four-year nonresearch colleges and universities. Little research has been done to look at how such programs should be designed to meet the needs of all undergraduate students.

In this chapter, we discuss aspects of designing a successful and inclusive research experience—from recruiting and application to program components and evaluation. We end with an overview of three programs that were specifically designed to meet the unique needs of underrepresented populations. We are particularly focused on research programs related to the Anthropocene and how they might give students an understanding of critical issues that move from a local to a global scale—starting with the students' individual research projects and lived experiences and then relating to global issues, and finally bringing new understanding back to refocus on the local.

## 8.1. CONSIDERATIONS FOR PLANNING AN INCLUSIVE UNDERGRADUATE RESEARCH OPPORTUNITY PROGRAM

### 8.1.1. Goal of the Program

Some of the strongest programs currently available to students have clearly defined missions. Some of the main goals institutions have for offering undergraduate research programs include: recruiting students for the university's graduate programs, promoting the retention of current undergraduate students, supporting current research projects, creating long-term partnerships between the university and other institutions (e.g., two-year colleges, minority-serving institutions [MSIs] or four-year colleges) in support of long-term recruiting strategies, promoting diversity within the sciences, increasing the number of students in STEM programs, increasing the excellence of current majors at the university, and creating or strengthening collaboration with a community to work on local pressing issues. If a program's goals include recruiting nontraditional students, faculty designing the program need to become familiar with obstacles that most REU programs present to nontraditional students: spouses, children, jobs, aging parents, checkered academic careers, and so on.

### 8.1.2. Identify the Ideal Participant

Once clear goals are set for starting an undergraduate research program, it is possible to begin identifying the potential participants. Some programs may gear themselves toward first- or second-year students, such as students from community colleges (CCs) or MSIs. Others may be geared toward rising seniors, who are fairly settled into their academic careers and may be excellent recruits for graduate programs. Programs that have need for students with advanced mathematics or physics abilities, for example, may, out of necessity, be aimed at rising seniors. On the other hand, programs that aim to promote diversity in the sciences, support retention, or increase the number of students entering STEM programs may be interested in creating a program that allows for rising freshman or sophomores as participants. This enables students to experience research at an early point in their careers, which is known to have a positive impact on students' choosing STEM careers [*Russell, et al.*, 2007; *Thurgood et al.*, 2010]. A program can also focus on participants who are struggling some with foundational material, such as calculus or physics, with the assumption that these students can be counseled to focus on fixing these issues once they are motivated by participation in research projects.

### 8.1.3. Advertising and Recruiting

Advertising and recruiting for research programs follows directly from decisions about the type of participants. With the plethora of opportunities afforded to undergraduate students, programs need to be proactive in recruiting. Making use of e-mail listservs, professional societies' contacts, and trusted colleagues should enhance the applicant pool. A well-planned research program will have a strategy for widening and materials for enticing a large potential pool of participants. Programs that directly address anthropocenic issues, such as global climate change, migration of species, changes to the oceans, and so on, will all have a particular challenge in recruiting for diversity. A recent survey of NSF's REU geoscience programs for 2009 to 2012 reveals that a majority (~ 90 percent) of participants are Caucasian [*Rom et al.*, 2012]. This condition persists despite decades of attention by NSF on promoting diversity in REUs. We will focus the discussion that follows on recruiting for diversity, as well as designing programs that allow for broader participation.

### 8.1.4. Recruiting for Diversity

Strategies for program design that support participation by students from MSIs and CCs take into account that many students from these institutions are considered nontraditional. Recent statistics of CC students show that the average age is 29, 59 percent are part-time, and 27 percent are parents [*American Association of Community Colleges*, 2013]. Research experience programs have historically attracted the traditional (nonmarried, young, full-time) student from a four-year institution. Recent data show that NSF REU participants overwhelming come from institutions granting doctorates, although there is minimal to no participation from historically black colleges and universities (HBCUs), CCs, or tribal colleges [*Rom et al.*, 2012]. In general, research is focused on the need for supporting faculty research at these institutions (and thus increasing student participation in research) or simply improving STEM students' persistence rates, rather than increasing recruitment of these students into current REU programs [i.e., *Perez*, 2003; *NRC*, 2012b]. Thus, unfortunately, it is not known if these potential participants do not know about the programs, they know but do not apply, or they apply but are not selected to participate. The authors' experience is that students often do not know about these programs, but with encouragement, the students are excited to apply, especially if programs are designed with the nontraditional student in mind. However, lack of coursework in higher-level math, chemistry, and physics programs may still be a barrier for entry into some REU programs.

The authors have found success in recruiting for diversity for their programs and can suggest the following strategies:

1. Concentrate efforts on building partnerships with a few key institutions that have shown promise as recruiting grounds. Get to personally know faculty and advisors within these institutions, and if possible, visit the schools several times per year. Assure faculty and advisors that you will strongly consider any applicant that they recommend and that you will personally see to it that their students have a good experience in your program. Work on building and maintaining these relationships, and consider the faculty and advisors as experts who can give you further information about making your program sensitive to the needs of their students. Identifying strong programs at two- and four-year colleges is a useful strategy for focused recruiting and can also lead to the establishment of long-term relationships that support future recruiting efforts. NSF's WEBCASPAR site (www.webcaspar.gov) is an important tool for identifying these programs and can also be used to find useful discipline-specific statistics related to student enrollment and graduation at institutions across the United States.

2. Build a database of contacts who work with underrepresented students and friends of your program to help you recruit. Connect with past participants and the faculty and advisors who recommended past participants. Attend minority professional conferences, such as the Society for the Advancement of Chicanos and Native Americans in Science or the American Indian Science and Engineering Society, and while you are there, work to meet faculty and advisors as well as students. Send them updates about your program and several times throughout the recruiting period, contact them via e-mail to remind them about your upcoming deadlines. Consider setting up a system in which a person could be nominated to participate in your program.

3. Work with groups such as the Institute for Broadening Participation, which work to get information about research opportunities out to students from groups underrepresented in the sciences. Also make connections with programs in your own institution and others that mentor these students, such as the Louis Stokes Alliance for Minority Participation programs and the McNair programs.

4. Build a reputation as a diverse program, and you will naturally begin attracting students. Your pool of students from groups underrepresented in STEM fields will increase with time if you show a commitment to having a

diverse group, and you will increasingly attract exactly the candidates you are targeting.

5. Recruit in groups with other related programs and create a supportive network of recruiters that work together to attract students and place them in the program that is right for each student.

6. Recruit in cohorts. If the students have to leave their institutions/communities to participate in an REU, then designing an enrollment strategy that allows for multiple students from the same college or community will create a safe, comfortable learning environment away from home. A program could also bring together a supportive group of students with mixed backgrounds to develop a cohort, even if they are from different schools or communities.

It is useful to have recruiting and application materials reviewed by advisors from McNair, Trio, or other minority programs to see if the materials are inviting, inclusive, and specific about required qualifications without being intimidating. Recruiting materials should accurately reflect the expectations of your participants and outline dependent-friendly policies.

### 8.1.5. Attracting Students with Disabilities

Another often overlooked aspect of diversifying the geosciences is welcoming and retaining students with disabilities. The participation of persons with disabilities in the geosciences is lower than in the STEM workforce and in the overall US population. Citing 2000 US Department of Labor statistics and NSF data, *Locke* [2005] reports that, although of persons aged 16 to 64, 18.6 percent have a disability, persons with disabilities account for only 7 percent of the science and engineering workforce and only 2.4 percent of the doctorate recipients in the geosciences reported having a disability. Activities generally perceived as essential in the geosciences, such as field trips, instrumentation work, or image and graphics interpretation can be perceived as barriers for students with different mobile, visual, or audio impairments. Still, great strides are being made by innovative educators to counter these concerns [*Asher*, 2001; *Klemm et al.*, 2001]. Research internships can serve on the forefront of innovation by actively recruiting and supporting students with disabilities into their programs. Organizations specific to our field, such as the International Association for Geoscience Diversity (http://www.theiagd.org/), can assist faculty and administrators to design accessible research experiences.

### 8.1.6. The Application

Program applications are an important tool for communicating the goals of the program to both draw applicants and guide potential participants to share pertinent information about themselves. The questions on the application predetermine criteria that reviewers will use to select participants and need to be clearly aligned to programmatic goals. The questions inform applicants about the program and create an image of the ideal applicant. Will that image mirror the candidate the program hopes to attract? If not, the application may discourage potentially great students from applying.

The personal essay can be a barrier for some applicants. Some undergraduates have already had experience writing personal essays as part of the college application process, but for some, this could be their first experience with this task. One strategy is to break the essay down into a series of short-answer questions as an effective method of supporting students to give excellent, to-the-point information about themselves, specific to the program's needs. Some examples include:

- "Tell us why you want to participate in this research experience."
- "Please describe your academic and professional goals in both the short term (next few years) and long term (10 years out)."
- "Describe how participating in this REU could assist you in achieving these goals."
- "Describe what attributes you bring to this program that will make you an asset to your research team."

### 8.1.7. The Selection

Choosing your student participants is key to a successful program. To create a clear selection process, take the time to outline the process and criteria, including who decides who gets admitted. All decisions should follow the goals and purpose of the program. The ultimate goal does not necessarily need to be graduate school because that is the old line of thinking. Redefining success of undergraduate research programs to include the impact participants can have in many economic sectors (i.e., as teachers, technicians, resource managers, or research assistants) makes sense and may allow for more inclusive admittance criteria. Many underrepresented and nontraditional students have not been exposed to the myriad STEM areas. So one goal of an undergraduate research program (especially if working with students from CCs and MSIs) could be to allow the participants to explore STEM through research. In this case, the ultimate goal may be to have the participant matriculate to a four-year institution to earn a bachelor's degree in a STEM field.

### 8.1.8. Supporting Nontraditional Students

As noted previously, one way to increase participation by a broad, diverse group of students is to make your program friendly to the nontraditional student. Students from underrepresented groups are more likely to be nontraditional students, attend CCs before moving on to four-year programs, and to support dependent children [*Horn*, 1996; *NRC*, 2012b]. Thus, recruiting students from CCs, students who are nontraditional, and parents with children is one way to tap into a pool of students who rarely participate in these types of programs. The added maturity of these students can sometimes be a real benefit to the program; however program managers need to be concerned that students not be overburdened with both a full-time research position and providing dependent care in an unfamiliar environment (away from home, the familiar daycare facilities, schools, etc.). Some thoughts on making a program parent-friendly include: provide housing that can accommodate families, provide funding or support for dependent care for students with children or dependent adults, consider the hours students work and make them flexible, and be sensitive to scheduling social activities for the cohort because early morning and evening activities could conflict with dependent care or parenting responsibilities. In addition, to avoid some of the issues that come with having parents as participants, locate programs in the students' home community and allow for commuter participants rather than having them exclusively resident based. This gives programs the opportunity to recruit from nearby CCs or institutions that are not research intensive.

Programs that account for some of these attributes in the design phase may be able to create support structures that other programs are not able to offer or add in. Particularly through nurturing these students as underclassmen, there is a real potential that these will become future recruits for graduate programs.

### 8.1.9. The Orientation and Final Week

A good orientation and a good final program can bookend your undergraduate research program and provide students a framework for understanding the point of all their work. Participants and mentors alike need to have a firm grasp of the expectations of the program assignments, deadlines, and after-summer commitments/opportunities (e.g., additional data acquisition, travel to professional meetings to present a poster, and so on). Equally important is to clarify what the program provides (i.e., stipends, transportation, housing, food, and such) and what might not be provided (i.e., extra baggage fees, lost key replacement, and so on). Communication needs to be fostered through meetings, collaborative technologies, and group events. Evaluation methods for the program need to be developed and shared with participants and mentors. (See Chapter 7 for an informative discussion of culturally specific evaluation practices).

Beyond the nuts and bolts, however, the orientation and final sessions can be a driver for communicating the larger vision, mission, and objectives of the program. Discussions can be fostered on the goals of the research project(s), the backgrounds of the mentor team, the backgrounds of the participants, and how these fit into larger issues, such as promoting better understanding of how the research fits within the discipline, the larger scientific community, and the global community. If these are well framed at the beginning of the project and revisited at the end of the project, they can add meaning for both participants and mentors. Families of participants and mentors can often be profitably included in orientation and final week activities through events such as picnics or poster sessions. This can help build esprit-de-corps and allows participants and mentors a chance to share noncollegiate aspects of their lives, while building family support for the research project.

### 8.1.10. The Devil Is in the Details

It's been said "the Devil is in the details" and this is certainly the case for summer research programs. Because participants and mentors need to accomplish a lot of work in a short period of time, planning, organization, communication, and implementation make or break a program. For project leaders, the goal is to develop a pathway for success for their participants. Before beginning the program, participants need to know about housing, stipends, insurance, transportation, what to bring, what's provided, and how to prepare themselves. Research plans, field trips, social activities, closing week, and evaluation activities should be designed, detailed, and on the calendar before the program starts. Mentors and participants need a clear roadmap of expectations for both their own role in the program and each others. Some programs start with a clearly defined research question and plan for execution, others have students develop the research plan. In either case, by the end of the orientation week students should be engaged in the project and have a clear idea of how they will proceed. Programs often begin by having students do background reading (either before arriving on campus or during the first week), but it is essential to offer students the opportunity to get hands-on, even during the orientation, to promote engagement.

Communication between and among the mentors is extremely important, as well as clear communication with the students. Set up opportunities to find out what the students are thinking and where they are having issues: a weekly short reflection paper, a weekly meeting, online discussions on Facebook, or a course management system. This is also where there is an advantage to having a team of mentors—ideally someone besides, or in addition to, the research mentor will be asking these questions of the students to identify barriers to communication between the student and their primary research mentor. Plan opportunities to enhance socialization between students and mentors. It may be beneficial to plan activities with other undergraduate research programs either in-person or via videoconference. Always remember that these are undergraduates on a tight timeline. Encourage, and if necessary, help students to do the writing and poster development from the first week on. Finally, as much as possible, plan for the unexpected. Students who are away from home, perhaps for the first time, need to have a safety net in place in case something goes awry.

### 8.1.11. Multiple Advisors and Team-Structured Collaborative Research

Increasingly, students in undergraduate research programs are being placed on research teams rather than pairing an individual student with an individual faculty member. Both methods of running a research experience program have positive and negative factors. When students are paired individually with a faculty member, under ideal conditions, the potential for meaningful interaction between the faculty member and the student creates learning opportunities that far outstrip what a student can gain in the classroom. This interaction is highly dependent on the time and energy the faculty member and student are willing to devote and the relationship between the two. In the authors' experience, these relationships can lead to student success over the course of the research project, to student failure at completing the project, or everything in between. There is value in having a team of mentors, as opposed to one advisor, so students have a variety of people with whom to interact. Mentors can also be at different stages—faculty, post doctorate, grad student, or peer. If a program allows for it, multiple mentors can address separate issues with the student. Some programs have found it useful to assign separate research, computing, and writing mentors to a student. The workload is then balanced out over several people and a student has the benefit of multiple perspectives and mentoring styles.

One primary motivation for placing students on teams is to create a more supportive environment. A single undergraduate student on a research team can be led to feel isolated or that they are the only one who does not understand things and are holding back the research. Having the support and companionship of at least one other undergraduate can help to dispel these feelings. In addition, research on collaborative learning indicates positive enhancements to learning from teams: students can learn an enormous amount from each other [*Smith*, 1991; *Johnson et al.*, 1998; *Gates et al.*, 1999]. Students can be encouraged to work minor problems out among themselves before they bring them to their research mentor, thus facilitating independence, fostering inquiry, and minimizing the stress of too-frequent interruptions to the research mentor. If the advisor assigns individual parts of the project to each participant, but encourages them to work as a team, the members can support one another in understanding the related scientific literature, compiling the project background, and developing their methodologies. These individual parts should stand on their own while contributing to a larger effort. This also allows participants to put their own narrow project into a wider perspective. Furthermore, conducting research in cohort-groups helps to maintain the community feel, which could create a supportive environment for students from minority groups, especially if the cohort is created carefully to provide a mix of backgrounds and experiences.

### 8.1.12. Scaffolding Research Experiences

Scaffolding undergraduate research projects allows students to participate over several years so that they can become experts and peer mentors. These senior students are assigned more responsibility, make more money for their efforts, and are afforded more benefits (like presenting at a national conference, working on a more complex project). Scaffolding undergraduate research programs could allow for students to participate first in a project in their home community, become familiar with faculty mentors and research, then matriculate to a large research university's summer program in subsequent years. By first participating in a local research project, the unknowns of research have been eliminated in a safe environment and the students are more confident in their skills.

Alternatively, scaffolding could be accomplished by incorporating a two-week mini-research experience into the summer-long program. This mini-program could be open to late-high-school and early community college or tribal college students to introduce the experience and eliminate some of the unknowns. These students would then be invited to apply for the full program the following year.

Finally, by collaborating with other programs, scaffolding can occur as students move through a variety of experiences. Programs that accept rising freshmen and sophomores can send them along to a partner program if they have shown promise in their first research project. Mentoring shared between partner programs can assure the student continuous support as they move towards graduation.

## 8.2. TEACHING STUDENTS ABOUT THE ANTHROPOCENE IN RESEARCH EXPERIENCES

### 8.2.1. Ethics Training and the Anthropocene

It is difficult to incorporate anthropocenic issues into a summer research program. Geoscience concepts related to the Anthropocene are by their nature complex and global. In addition, participants come with a wide range of background on climate change issues, considering that some of their home institutions do not offer formal course work covering the Anthropocene or even present a climate skeptic perspective to their students. Undergraduate summer research must, of necessity, be narrowly and well-defined, and simplified if the projects are to be completed within the course of a summer. However, participants can be led to making the complex relational connections that support understanding of anthropocenic issues through a well-structured ethics component within the program. A new approach to ethics training asks students to understand their research activities from both a local and global perspective as a means to better understanding their role as future researchers, moving beyond standard research ethics. A group discussion that begins during the orientation and continues throughout the summer can encourage students to occasionally look up from their microscope or sample to consider their research in a global context. Almost all undergraduate research programs have at least a minimal discussion of research ethics that incorporates topics such as authorship, research projects that involve humans or animals, plagiarism, misrepresentation of data, and so on. It is possible for anthropocenic-related undergraduate research programs to broaden that conversation to a frank discussion of the impact of humanity on the planet and its species. Helping students to relate their own research projects to these larger global questions can bring added meaning to their research. This discussion can include an exploration of the impact the research itself might have on the planet— what is the carbon footprint of the program, what resources are being corralled in service of knowledge, what efforts are being made to ensure minimal adverse impact to the environment from the tools and materials used in the research project? How will this be balanced by the contribution the research will make to supporting sustainability? By fostering mindfulness about the research, and both the positive and negative possible impacts of the research project on the planet, students gain a better understanding of the Anthropocene, of sustainability, of research, and the role they play. Participants can also be asked to relate what they are doing to their home community—what do they bring to the team from their background, and what will they take home with them. These types of questions take the student beyond standard research ethics, and ask the participant to consider their own role as a scientist in the larger community.

The value of rethinking standard research ethics training to focus more broadly has been subject of recent scholarship. For example, *McGinn and Bosacki* [2004] address the issue in relation to graduate education, but it is as pertinent for undergraduate research programs:

> We suggest that research courses need to provide course space to encourage students to question ethical guidelines, and to explore and discuss the complex moral and ethical issues that underlie research methods. We aim to encourage students to become wise researchers who use both their minds and their hearts when making research decisions. Research courses need to encourage students to develop competencies as researchers who will reflect upon their research decisions with respect to pragmatic, philosophical, ethical, and moral issues related to research design, methods, data analysis, and presentation [*McGinn and Bosacki*, 2004].

> It makes people think—what is this world about? How can I help? More people on the reservation need to look beyond what they are used to seeing—[summer research programs] can do that.—REU participant

### 8.2.2. Creating a Place-Based Undergraduate Research Program and Incorporating Community-Based Research Projects related to Anthropocenic Issues— Working *with* Communities, Not Just *in* Communities

Another powerful way to relate issues of the Anthropocene to students is through place-based research programs. Research on place-based learning supports the idea that if students are involved in a research project that is focused on their home community, on a topic that is of significance to their personal lives, then the students will be more engaged and learning will be more successful [*van der Hoeven Kraft*, 2011]. In addition, students can bring relevant knowledge to the project of which even the project leader may be unaware. Finally, the undergraduate researchers can be introduced to the ethics of doing research in and with a community both from the perspective of a researcher and of a community member. For these reasons, undergraduate research programs that are set up with and within a community and draw their participants at least partially from that community can have a deep impact on the participants. Researchers who wish to set up a program of this type must have a firm grasp of the principles of community-based participatory research themselves to grapple with the issues their undergraduate participants may experience as part of a research team that is located within a community [i.e., see *Mariella et al.*, 2009]. The rewards of this type of program can be huge, however, particularly if community members become part of the audience who are interested in the outcomes of the research project (or even become involved in the project themselves, perhaps offering traditional or local knowledge to the research team).

The research program should be driven by the needs of the community and not the needs of the organizer or researcher. As part of the relationship building, the research team and community representatives should work alongside one another to develop the research agenda. As an example, key to success in tribal communities is the development of a mutually beneficial relationship built on a strong foundation of reciprocated trust and common goals. Relationships are built over time with the responsibility of community building on the side of the organizer or research team. Time must be invested up front to ensure that all parties share common goals, understand the intent of the program and the expectations for the participants. For a more thorough discussion of this topic, see Chapter 6.

## 8.3. THREE UNDERGRADUATE RESEARCH EXPERIENCES RELATED TO THE ANTHROPOCENE

### 8.3.1. The Significant Opportunities in Atmospheric Research and Science Program

One of the challenges of teaching students about the Anthropocene through research internships is the fact that these experiences only last 7 to 10 weeks. During this short time, students are only able to work on a small piece of a large puzzle, and it can be challenging to teach both research methods with a focus on small details, and the understanding of large complex systems such as the changing global climate. So how can we provide a successful personalized research experience, ensure the students conclude the summer with results to be presented at national conferences, and at the same time, teach the students to ground their own work in the larger picture?

One way to expose undergraduate students to the study of the Anthropocene is by giving them direct access to the latest research. In the Significant Opportunities in Atmospheric Research and Science Program (SOARS), students (called *protégés*) get to work one-on-one with some of the leading researchers in the field. SOARS is an internship and mentoring program with a mission to increase the number of students from historically underrepresented groups who enroll and succeed in graduate programs in the atmospheric and related sciences. The program is managed by the University Corporation for Atmospheric Research (UCAR) and hosted at the National Center for Atmospheric Research (NCAR) in Boulder, Colorado. SOARS protégés work at NCAR, NOAA, the Cooperative Institute for Research in Environmental Science (CIRES), the University of Colorado at Boulder, and the NSF-funded Center for Multi-scale Modeling of Atmospheric Processes (CMMAP) at Colorado State University. Topics of research span the broad field of climate and weather, including computing and engineering in support of the atmospheric sciences.

SOARS accepts students from most majors, so there are a significant portion of students who never had any course work in climatology or climate change impacts. Their research interests being more of a driver than their academic background, about a third of protégés each year have the opportunity to work on climate modeling, climate change impacts, or adaptation topics during the summer. Other research topics cover meteorology, scientific engineering, or computing. Those students who might not directly work with a climate scientist can get exposure to the topic through colloquia and presentations given at NCAR and through tours of the supercomputers and the climate modeling divisions.

Another way for students to learn about the Anthropocene in SOARS is through peer learning. The SOARS program provides a weekly communication workshop that includes a peer-review process. Students not only help each other improve their writing style and grammar, but also learn from one another's research topics. They start to see connections between each other's projects, and even as they work on their small piece of the puzzle, start to appreciate the bigger picture. Protégés also participate in weekly ethics discussions. Like many other programs, protégés discuss topics such as plagiarism, human subject research, and co-authorship. Discussions also cover their role as a scientist, ethics of research, and the merit of advocacy in science. Another way to foster discussion is through community outreach; each protégé is required to participate in at least one public outreach activity during the summer. They can chose from media training and being available to the (local) press, doing hands-on science outreach with K–12 groups visiting NCAR, or blogging about their project or other science-related issues. The program offers the protégés presentation training and opportunities for practicing elevator speeches about their work so they get comfortable speaking about their research to a wider audience. Through training, protégés understand the context in which their own project is embedded and practice answering the question "Why does your research matter"? Engaging in these outreach activities spurs intense discussion about the role of a scientists and addresses some of the questions the public has for geoscientists, often about climate change and intensification of severe weather events.

One way the short-lived nature of the research internship has been addressed, is by allowing SOARS protégés to participate in the program for up to four summers. This allows them to explore the breadth of atmospheric sciences and offers additional support as they transition into graduate school. Returning multiple years allows protégés to reflect on their work and pursue a new line of inquiry every summer or to dig deeper into a topic they started the prior summer. Returning protégés have a key part in defining their own research projects and often the topics lead to graduate thesis work.

How does the study of the Anthropocene relate to the SOARS mission? Studies have shown that students from traditionally underrepresented groups shy away from the geosciences because they do not see it as a field that has a direct positive impact on their communities [*Seymour and Hewitt*, 1997; *Bembry et al.*, 2000]. Protégés come to SOARS with a desire to work on projects addressing these issues. In fact, some propose projects and get the support from SOARS to follow their own line of inquiry, informing and influencing NCAR research by doing so (see https://www2.ucar.edu/atmosnews/research/7721/mapping-hurricane-vulnerability).

Some protégés even go a step further and do field work to learn more about climate change impacts already felt by local communities. A new partnership with the Center for Hazards Assessment, Response and Technology (CHART) at the University of New Orleans, the South Louisiana Wetland Discovery Center, and local Native American and Cajun communities in southern Louisiana, has allowed SOARS protégés to study community viability in the face of environmental change. Protégés work directly with the communities to define research problems and integrate physical science, geospatial technology, and traditional ecological knowledge. These off-site research opportunities are limited to advanced protégés who have experienced the support from the SOARS cohort during their first summer in Boulder and can sustain the contact through social media and weekly phone calls while being in the field.

Increasing the participation of students from traditionally underrepresented groups in the geosciences will have a larger success if we invite students from these communities to bring their own research questions to the program. Instead of prescribing a project, opening up a dialogue about what the student is curious about will greatly enhance the student's experience, increase the chances he or she will be retained in the geosciences, and has the potential to inform our sciences [*Sloan and Haacker-Santos*, 2012].

The SOARS Program is managed by the University Corporation for Atmospheric Research and funded by the National Science Foundation (NSF-0618847), the National Center for Atmospheric Research, the University of Colorado at Boulder, and by the Center for Multiscale Modeling of Atmospheric Processes.

### 8.3.2. The Research Experience for Undergraduates on Sustainable Land and Water Resources

The Research Experience for Undergraduates on Sustainable Land and Water Resources grows out of the shared goals of partners—Fond du Lac Tribal and Community College (FDLTCC); the Fond du Lac Band of Lake Superior Chippewa in northern Minnesota (FDL); Salish Kootenai College (SKC), a tribal college located on the Flathead Reservation in Pablo, Montana; and two research centers at the University of Minnesota—the National Lacustrine (lake) Core Facility (LacCore), Department of Earth Sciences, and the National Center for Earth-surface Dynamics (NCED), headquartered at the St. Anthony Falls Laboratory (SAFL). Together they are striving to bring research opportunities to undergraduate students from across the nation, with an emphasis on providing pathways into research opportunities for native students; create diverse and supportive research teams that bring together students from different disciplines, academic institutions, cultural backgrounds, and life experiences; construct pathways for students from two-year programs into four-year programs, and from four-year programs into graduate programs that further the academic and career interests of the participants; develop new team mentoring structures that support mentors as they work to support the participants; center research on sustainability issues that are impacting native reservations, especially those related to anthropocenic change; develop projects that are place-based, community-oriented, incorporate the local native culture, are led by reservation resource managers and tribal college faculty; and develop the next generation of land and water resource managers.

Research projects in the REU on Sustainable Land and Water Resources are focused on topics that integrate Earth-surface dynamics, geology, ecology, limnology and paleolimnology, and hydrology. Research projects focus on the principles and techniques required for sustainable land and water management and restoration while using a team-oriented approach that emphasizes quantitative and predictive methods. Team Stream focuses on stream restoration related experimental work at St. Anthony Falls Laboratory and is led by faculty from NCED; Team SPAW examines the effects of surface water management practice on groundwater resources in Montana and other projects related to local resources and is led by faculty from the Hydrology program at SKC; and Team Zaaga'igan ("lake" in Ojibwe) investigates the paleolimnology and current habitat conditions of lakes, rivers, and bogs in northern Minnesota, led by faculty and researchers from LacCore, FDL, and FDLTCC.

A major challenge of the program is to knit together teams into a cohesive group while running teams in three geographically separated locations (Minneapolis and Cloquet, Minnesota; and Pablo, Montana). Weekly all-team videoconferences help students get acquainted, and everyone is brought together into an all-team gathering at the end of the summer. Each week, one of the teams leads the videoconference, presenting details about their research projects. This approach was developed at NCED to unify researchers from NCED institutions and further the center's research objectives, but the format proved easily adaptable to the needs of the REU, which are similar.

It is also important to build strong teams at each site. To support teambuilding, each team works on a single overarching problem, while individuals each have their own project that is a small piece of the overall project; students write portions of their paper together as a team (background, literature survey); each team has a name to support group identity and continuity from year to year; team-building activities are incorporated into orientation; and weekly team meetings are held.

The end-of-the-year all-team gathering is a unique and essential feature of the REU. Participants come together for 7 to 10 days at the end of the summer. Some of the activities of the all-team gathering include: research talks given by all participants; a joint poster session with NCED's Summer Institute on Earth-surface Dynamics, providing an excellent audience of graduate students and faculty for student posters; ethics discussions; "next move" discussions where participants are able to talk with faculty, graduate students, and other mentors about their future academic and life plans and concerns; a field trip to a geologically interesting site in Minnesota; and multiple opportunities for socialization: picnic, bar-b-que at SAFL, Twins baseball game, and awards dinner.

Two of the teams are located on American Indian reservations, which allow for participation by local students who are attending SKC, FDLTCC, or other nearby tribal colleges. This has had the result of drawing undergraduate participants who have dependent children—in fact during the first year the rate of participation by parents was at 60 percent of participants, a surprising result that was not anticipated. The main tension that has arisen out of this situation has been in keeping the grant management consistent with federal law and institutional policy while allowing for the inclusion of these family members at the REU events and activities. Because the team has long experience working in American Indian communities, where inclusion of family members is consistent with cultural traditions, mentors were comfortable in allowing for enough flexibility in the program to support parents as participants.

In many ways, the REU on Sustainable Land and Water Resources is a unique experiment in finding ways to create diverse teams that support broadened participation in the geosciences, while also supporting community-based participatory research on American Indian reservations. The orientation and end-of-the-year ethics discussions are focused on asking participants and mentors to think about what they bring to the program and what they will take back to their communities from the program.

The REU on Sustainable Land and Water Resources is made possible with the support of the National Science Foundation (EAR-1156984) and the National Center for Earth-surface Dynamics (EAR-0120914).

### 8.3.3. The GEMscholar Program: An Innovative Research Program for American Indian Undergraduate Students

The GEMscholar (Geology, Environmental Science, and Meteorology scholars) program seeks to increase the number of American Indian students pursuing graduate degrees in the geosciences—an area of higher education that has previously been unaddressed possibly because of low numbers of American Indian students holding baccalaureate degrees in geosciences. GEMscholars was intentionally designed to provide opportunities for the participants to experience hands-on science in their local settings. American Indian student education models were adapted focusing on the key themes of mentoring, culturally relevant valuations of geosciences and necessary career paths, and connections to community and family.

The model for recruiting, retaining, and graduating American Indians from undergraduate institutions and matriculating to graduate programs was developed based on a critical linkage between higher education and the tribal community's environmental and economic development. Through personal visits and relationship building over several months, a mutually beneficial partnership developed among a large research-intensive institution (RI), a smaller state institution (SI), and two two-year tribal colleges (TC). These partnerships were integral to the success and continuity of the program. To build a successful program, a strong foundation must be built on trust: lines of communication must be open, dialogue must remain honest, and actions must be transparent. To keep the relationship moving forward, continuous communication is key.

The partners sought to address the lack of ethnic diversity in the geosciences by using pedagogically and culturally sensitive learning opportunities to motivate American Indian students. The intent was to develop a model program where the students would follow a path of progress from earning their associate degree at the tribal college, their bachelor of science (BS) at a local state university, and their graduate degree in the geosciences at a large research-intensive institution. Along the way, they would participate in the GEMscholar summer research program to gain the skills, tools, and confidence necessary to be successful at each level of their progression. The program encourages the participants to give back to their local communities as native scientists and pass their newly gained knowledge to the others that come after them.

GEMscholar participants were recruited using various techniques. Former participants and program mentors circulated brochures and gave presentations at the TCs and SI. The most important allies in student recruiting proved to be the TC administration who would talk directly with students they felt would be most successful in the program. The former GEMscholars proved to be vital in recruiting students because they shared their positive experiences at their schools and in their communities.

The core strength of the GEMscholars program was bringing the summer research opportunity to the participants. The GEMscholars and faculty mentors (typically 12 participants, five faculty mentors, and one graduate student) met six times each summer in the participant's home community. Meetings and workshops were between three and six days to facilitate field and lab work. This project, studying the impact of invasive earthworms on the reservation's forests, was ultimately selected because the nature of the work (both field and lab components) would add to the student's skill set and the students could use the results of the work to directly impact their community. The program culminated in a trip to the RI where students presented their research findings at the GEMscholar Symposium. The product of the GEMscholars' research efforts also resulted in AGU conference presentations and publishable research.

Making the geosciences research relevant to the students was instrumental to increased interest in science research and continued participation and commitment to the program. The students found added value from the research in that they could contribute in a positive way to their communities. The hands-on nature of the research also connected the students to their local surroundings and resulted in a richer experience that fostered deeper understanding of the geosciences concepts related to the field investigations.

Mentors and GEMscholars working alongside one another in the participant's home community was paramount to the success of the project because it led to a tight bond among all the participants. Furthermore, the research was formulated with the intent to train the American Indian students and return the expertise back to the tribes.

Bringing together faculty from the TCs, SI, and RI has facilitated multiple collaborations; these exchanges continue to be fostered resulting in new research relationships and increased learning opportunities at each institution and in the local communities. Ideal joint research projects addressed environmental issues that tied the active student back to their home community and thereby, kept the connection between community and family. All the research projects have focused on the changing landscape as a result in part of effects of our changing climate. An unforeseen benefit occurred when participants discussed the research with community elders. The elders then shared historical information about the land, allowing the participants to identify the changes that had occurred over time. This information was shared with the community and has led to discussions on restoration and preservation.

This multifaceted program has the potential for increasing the effective pipeline of American Indian students to the TC for a two-year degree, to a four-year college for an undergraduate degree, and then to a graduate degree–granting institution. The goal is for the participants to graduate with master and doctorate degrees in geoscience and return to their tribal community as researchers, educators, and earth resource experts. Consequently, the GEMscholar program employed a comprehensive approach using mentoring and experiential learning opportunities. Because the overarching goal of the program was the betterment of the students, the program was built on: providing culturally relevant undergraduate research experiences; facilitating multi-tiered mentoring opportunities; expanding students' career and educational possibilities; and providing financial support. At the same time the faculty mentors benefited from: learning about Ojibwe heritage and community structure; working with American Indian students in their home communities; and visiting local communities and learning about the environmental concerns pertinent to geosciences research.

> I plan to graduate with a degree in science and continue to get a masters degree. Before GEMs I never considered graduate school now it seems possible.—GEMscholar participant

The GEMscholar Program is made possible with the support of the National Science Foundation Geoscience Education (NSF-0608034) and the hospitality and generosity of the Red Lake Nation College and Leech Lake Tribal College.

## REFERENCES

American Association of Community Colleges (2013), *Data Points: Community College Enrollment*, retrieved on December 3, 2013, at http://www.aacc.nche.edu/AboutCC/Trends/Pages/enrollment.aspx.

Asher, P. M. (2001), Teaching an introductory physical geology course to a student with visual impairment, *JGE*, 49(2), 166–169.

Bembry, J., C. J. Walrath, J. Pegues, and S. Brown (2000), Project Talent Flow II: SEM field choices and field switching of Black and Hispanic Undergraduates, University of Maryland–Baltimore County, Baltimore, MD (supported by the Alfred P. Sloan Foundation, grant 98-6-16).

Brownell, J. E., and L. E. Swaner (2010), *Five High-Impact Practices: Research on Learning, Outcomes, Completion, and Quality*, Association of American Colleges and Universities, Washington, DC.

Council for Undergraduate Research (2009), *About CUR: Timeline*, retrieved on December 3, 2013, at http://www.cur.org/about_cur/history/timeline/.

Crowe, M., and D. Brakke, (2008), Assessing the impact of undergraduate-research experiences on students: An overview of current literature, *CUR Quarterly*, 28(4), 43–50, retrieved December 3, 2013, from http://www.cur.org/assets/1/7/summer08CroweBrakke.pdf/.

Gates, A. Q., P. J. Teller, A. Bernat, N. Delgado, and C.K. Della-Piana (1999), Expanding participation in undergraduate research using the affinity group model. *J. Engrg. Ed.*, 88(4), 409–414.

Hancock, M. P., and S. H. Russel (2008), Research Experiences for Undergraduates (REU) in the Directorate for Engineering (ENG) 2003–2006 Participant Survey, prepared for The National Science Foundation, Directorate for Engineering, by SRI International, August 2008, retrieved April 10, 2013, from http://csted.sri.com/sites/default/files/reports/REU_ENG_2003-2006_Participant_Survey_Report.pdf.

Horn, L. J. (1996), Nontraditional undergraduates: Trends in enrollment from 1986 to 1992 and persistence and attainment among 1989–90 beginning postsecondary students, National Center for Education Statistics, NCES 97-578. Dennis Carroll, project officer, Washington, DC, 1996.

Johnson, D. W., R. T. Johnson, and K. A. Smith (1998), *Active Learning: Cooperation in the College Classroom* (2nd ed.), Interaction Book Company, Edina, MN.

Klemm, B., J. R. Skouge, R. Radtke, and J. R. Lazlo (2001), Ocean of potentiality: "Fully accessible" science camps, *J. Sci. Ed. Stud. Disab.*, 8, 22–29.

Locke, S. M. (2005), The Status of Persons with Disabilities in the Geosciences: White Paper, RAESM2 Symposium, October 13–14, Las Cruces, New Mexico, New Mexico State University.

Mariella, P., E. Brown, M. Carter, and V. Verri (2009), Tribally-driven participatory research: State of the practice and potential strategies for the future, J. Health Disparities Res. Pract., 3(2), Art. 4, retrieved from http://digitalscholarship.unlv.edu/jhdrp/vol3/iss2/4.

McGinn, M. K., and S. L. Bosacki (2004), Research ethics and practitioners: Concerns and strategies for novice researchers engaged in graduate education. *Forum Qualitative Sozialforschung/Forum: Qualitative Social Research*, retrieved November 14, 2013, from http://www.qualitative-research.net/index.php/fqs/article/view/615/1333.

National Research Council (NRC) (2010), *Informing an Effective Response to Climate Change*, The National Academies Press, Washington, DC.

National Research Council (NRC) (2012a), *New Research Opportunities in the Earth Sciences*, The National Academies Press, Washington, DC.

National Research Council (NRC) (2012b), *Community Colleges in the Evolving STEM Education Landscape: Summary of a Summit*, The National Academies Press, Washington, DC.

National Science Foundation (NSF) (1996), Geoscience Education: A Recommended Strategy, A Report Based on an August 29–30, 1996, Workshop from the Geoscience Education Working Group to the Advisory Committee for Geosciences and the Directorate for Geosciences of the National Science Foundation.

Perez, J. A. (2003), Undergraduate research at two-year colleges. *New Dir. Teach. Learn.*, 69–78. doi: 10.1002/tl.89.

Rom, E., L. Patino, J. Gonsalez, C. S. Weiler, L. Antel, Y. Colon, et al. (2012), An Analysis of NSF Geosciences Research Experience for Undergraduates Site Programs from 2002 to 2012, poster presented at the American Geophysical Union Fall Meeting, San Francisco, December 2012.

Russell, S. H., M. P. Hancock, and J. McCullough (2007), Benefits of undergraduate research experiences, *Science*, 316 (5824) (27 April), 548–549, retrieved November 14, 2013, from http://www.sciencemag.org/content/316/5824/548.summary.

Schowen, K. B. (2002), Value and Impact of Undergraduate Research in Chemistry: Lessons Gained From 45 Years of Experience at The University of Kansas, presented at CONFCHEM2002, Conference on Undergraduate Research; retrieved November 14, 2013, from http://www.files.chem.vt.edu/confchem/2002/a/schowen.html.

Seymour, E., and N. M. Hewitt (1997), *Talking about Leaving: Why Undergraduates Leave the Sciences*, Westview Press, Boulder.

Sloan, V., and R. Haacker-Santos (2012), Finding the Right Match: Pairing Undergraduate Research Interns and Scientists as a Way of Engaging Students in their Topic of Interest, 2012 GSA Annual Meeting in Charlotte.

Smith, K. A. (1999), Inquiry-Based Collaborative Learning, retrieved November 14, 2013, from http://www.nciia.net/proceed_01/smith%20handouts.pdf.

Thurgood, L., C. Ordowich, and P. Brown (2010), Research Experiences for Undergraduates (REU) in the Directorate for Engineering (ENG), Follow-up of FY 2006 Student Participants, Survey of Participant Experiences in NSF's Research Experiences for Undergraduates (REU) Program, Prepared for the National Science Foundation, Directorate for Engineering, by SRI International, August 2010, retrieved November 14, 2013, from http://csted.sri.com/content/research-experiences-undergraduates-reu-directorate-engineering-eng-follow-fy-2006-student-p.

van der Hoeven Kraft, K. J., L. Srogi, J. Husman, S. Semken, and M. Fuhrman (2011), Engaging students to learn through the affective domain: A new framework for teaching in the geosciences, *JGE*, 59(2), 71–84.

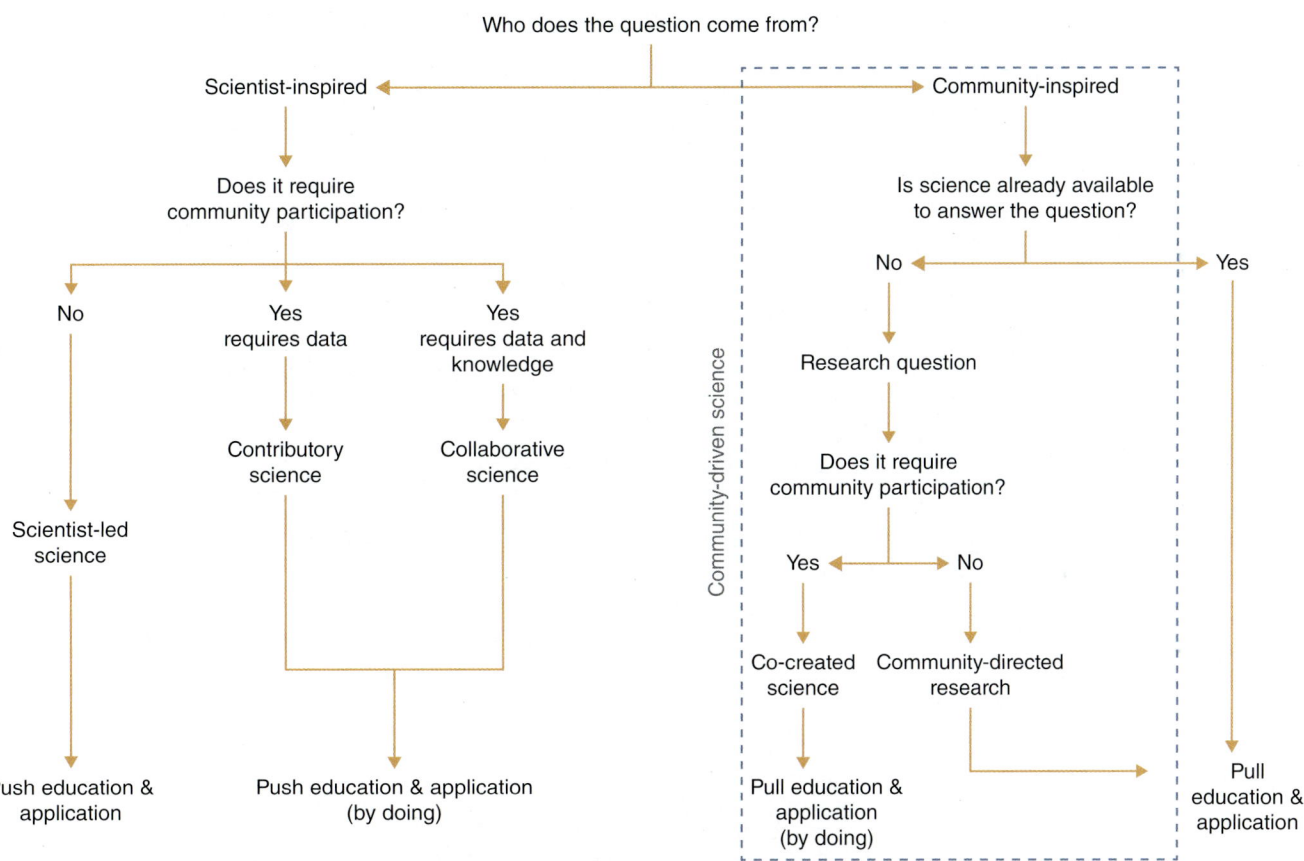

**Figure 6.1** A schematic tracing the different paths for connecting science and society. A key distinction is who participates in defining the scientific question, and this distinction flows into whether science results are pushed out from scientists or pulled into community priorities.

*Future Earth—Advancing Civic Understanding of the Anthropocene, Geophysical Monograph 203*. First Edition.
Edited by Diana Dalbotten, Gillian Roehrig, and Patrick Hamilton.
© 2014 American Geophysical Union. Published 2014 by John Wiley & Sons, Inc.

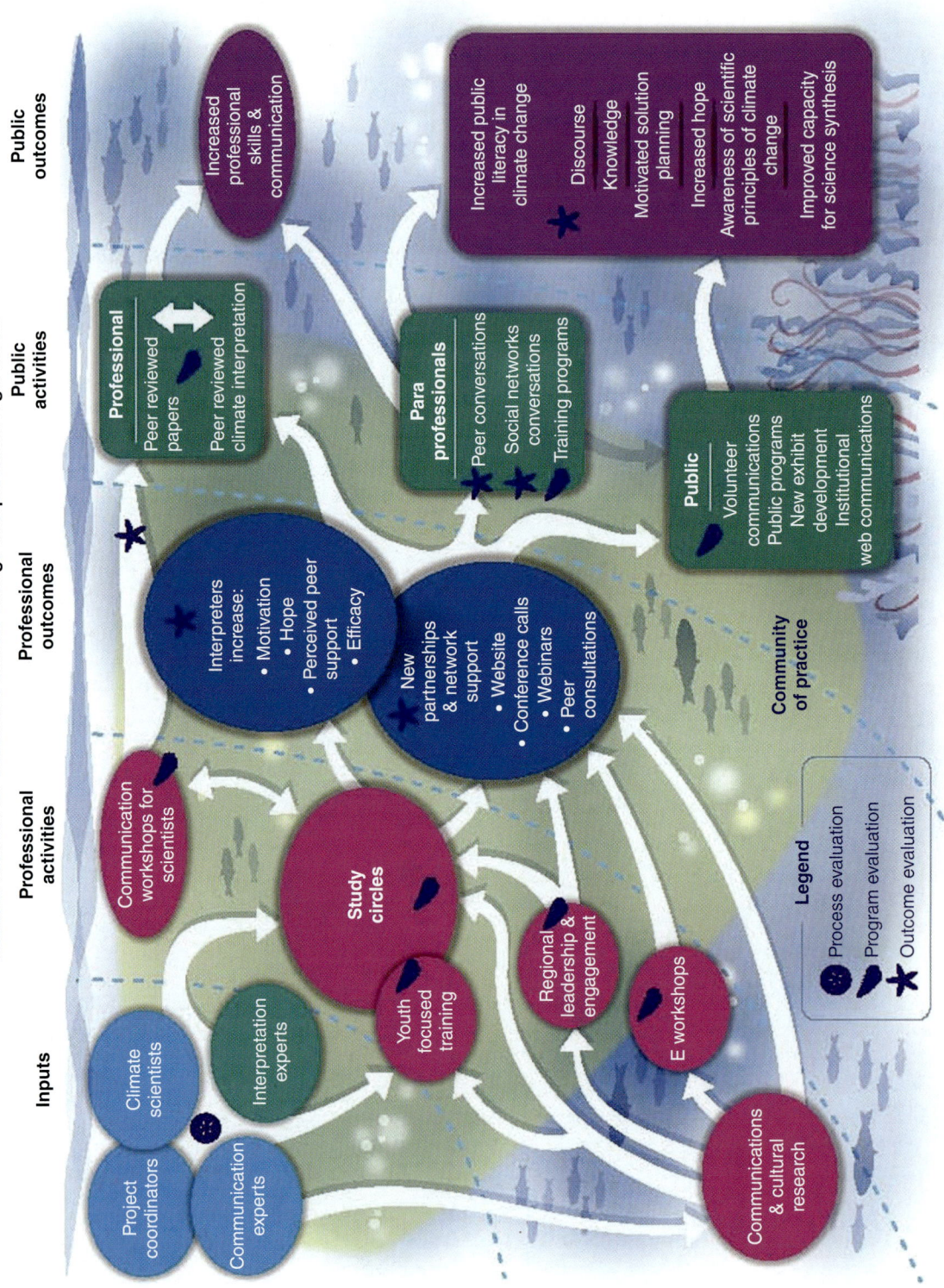

**Figure 9.1** NNOCCI logic model, showing the relationships among project inputs, activities and outcomes.

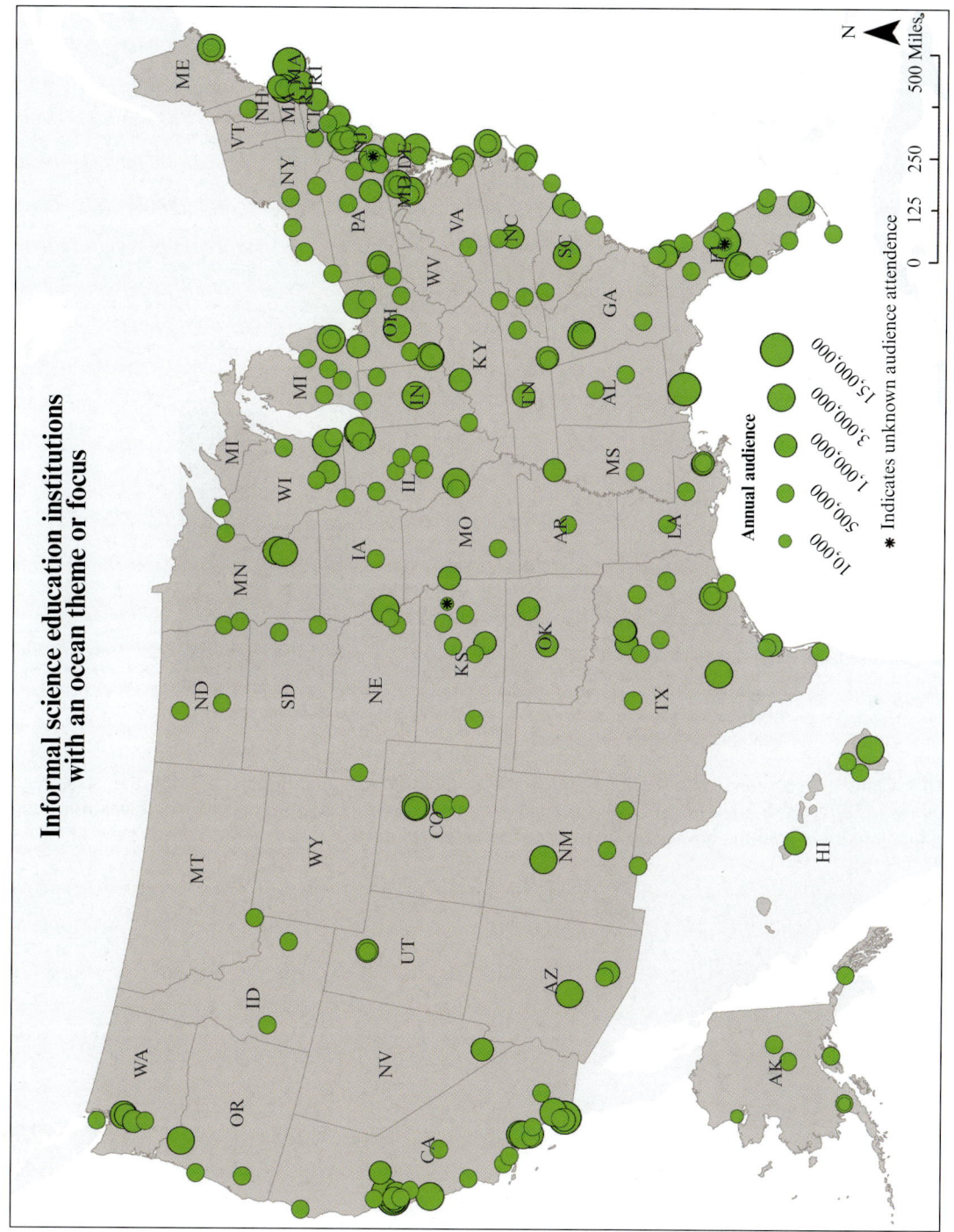

**Figure 9.2** NNOCCI has identified 212 ISEIs with an ocean theme or focus as potential participants; the goal is to reach 150 of these.

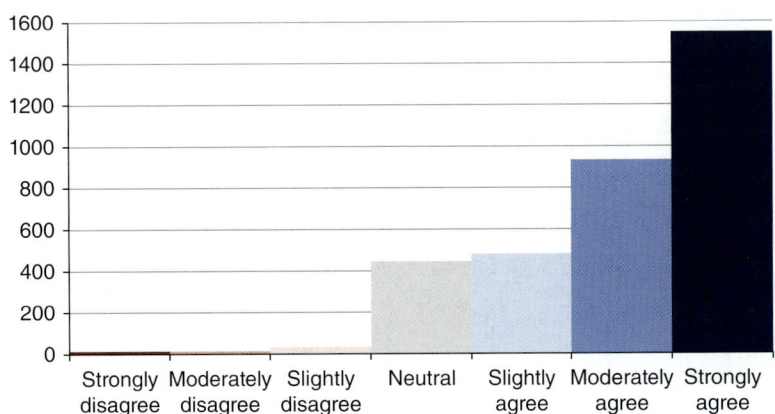

**Figure 10.1** Only 1.5 percent of aquarium and science museum visitors disagreed with the statement, *"Learning how to help conserve the ocean and its animals makes this a better place to visit,"* which a clear indication of the public's interest in learning how to be a part of conservation solutions.

# 9
# Shaping the Public Dialogue on Climate Change

William Spitzer*

## 9.1. PUBLIC UNDERSTANDING OF CLIMATE CHANGE

There is widespread agreement in the science community that climate change is adversely affecting marine environments [*Karl et al.*, 2009]:
- Effects of rising sea level and increasing storm intensity on natural habitats (beaches, salt marshes, barrier islands, and so on) and human-made infrastructure (e.g., homes, roads, bridges).
- Shifts in distribution, abundance, and productivity of marine species as a result of changes in water temperature, circulation patterns, and food availability.
- Ocean acidification and its impacts on shellfish, corals, and other groups of marine species because of carbon emissions.

The majority of the US population lives within 50 miles of a coast [*National Oceanic and Atmospheric Administration (NOAA)*, 2011], so a focus on coastal and marine impacts represents a promising means by which to help people advance their understanding of climate change through personally relevant impacts.

Among the general public, there is far less awareness of these impacts or of the critical role the ocean plays in the Earth's climate system [*The Ocean Project*, 2009]. Some people associate climate change with rising sea levels [*Lorenzoni et al.*, 2006], but most do not associate it with the other aforementioned changes in the oceans. Yet, understanding these issues is at the heart of both climate literacy [*NOAA*, 2009] and ocean literacy [*NOAA*, 2013].

### 9.1.1. Cognitive and Social Psychology of Climate Change

Public understanding of climate change lags far behind the consensus of the scientific community and not merely because the public lacks information about climate change [*Brechin*, 2003]. A recent psychological review identified several factors that make this topic particularly challenging [*Gifford*, 2011; *Weber and Stern*, 2011]:

1. **Inherent conceptual complexity**: Risk assessment is challenged by limitations of personal experience (e.g., overweighting recent events), inherent uncertainties in predicting behaviors of complex systems, and common misconceptions in mental models (e.g., seeing climate change as "pollution"). People tend to gravitate toward curtailment of negative behaviors, while overlooking the more powerful impacts of technical innovation and substitution [*Attari et al.*, 2010].

2. **Psychological factors**: Mental models can overly simplify climate change and, at worst, be incorrect. These models can lead to confirmation biases, resulting in reinforcement of current beliefs [*Shore*, 1996]. Further, affective and cognitive processes can lead to disengagement, avoidance, or feeling overwhelmed or disempowered.

3. **Social factors**: Climate change is framed as contentious. The news media tend to exacerbate this by aiming for "balance" even when there is an overwhelming scientific consensus [*Tannen*, 1999; *Corbett and Durfee*, 2004]. Public understanding is fragmented across political and ideological boundaries and has become more a question of belief than understanding [*Leiserowitz et al.*, 2008].

As a result, the prevalent "cultural models" available for understanding climate change are inadequate, leading to "ineffective personal actions and support for ineffective

---

*Vice President, Programs, Exhibits, and Planning, New England Aquarium, Boston, Massachusetts

---

*Future Earth—Advancing Civic Understanding of the Anthropocene, Geophysical Monograph 203*, First Edition.
Edited by Diana Dalbotten, Gillian Roehrig, and Patrick Hamilton.
© 2014 American Geophysical Union. Published 2014 by John Wiley & Sons, Inc.

policies, regardless of the level of personal commitment to environmental problems" [*Kempton et al.*, 1997, p. 220]. Fortunately, we now have a better understanding of learning and communication, based on advances in cognitive and social science research. Instead of simply conveying information, we need to facilitate "meaning-making"—helping individuals process information relative to their personal experiences and context.

A "strategic framing" approach to communication [*Bales and Gilliam*, 2004] supports meaning-making by building on careful empirical research to understand what people already value, believe, and understand, and then designing and testing communication strategies that help translate complex science in a way that allows people to examine evidence, make well-informed inferences, and embrace science-based solutions. This is a "nonpersuasive communication" strategy [*Fischhoff*, 2007] that entails explaining causes and consequences rather than advocating particular policies or actions.

This research-based approach can help to address conceptual, psychological, and social barriers described previously by creating new and more effective ways to recruit positive cultural models for understanding climate change. By providing scientifically accurate and well-tested metaphoric language, we can address conceptual complexity, overcome misconceptions, and demystify scientific processes. For example, the "greenhouse effect" is often used to explain global warming, despite the fact that most people do not understand how greenhouses work. An alternative and more effective metaphor is that carbon dioxide in the atmosphere creates a "heat-trapping blanket." As we burn fossil fuels in our daily lives, we release more carbon dioxide and thicken the blanket. And, as we know from our personal experience, a thicker blanket makes us warmer. If we want to reduce the warming, we need to find ways to reduce our consumption of fossil fuels. Research by the FrameWorks Institute has shown that when prompted with this metaphor, laypeople can better explain the links between fossil fuel use, increasing carbon dioxide in the atmosphere, changes in climate and the oceans, and possible solutions [*FrameWorks Institute*, 2010]. By understanding the chain of cause and effect, we can appreciate the nature of the problem, who is responsible, and what kinds of solutions are likely to be effective.

By focusing on specific applications and solutions to real-world problems, we can counteract "crisis" framing and despair. By appealing to strongly held universal values and concepts such as responsibility, stewardship, innovation, and interconnectedness, we can minimize polarization and contention. We must address the social context of climate change as well. A study of European museums presenting the topic of global warming [*Trautmann*, 2007, pp. 68–69] found programs and exhibits have "the best chance of inspiring changes in visitor understanding and behavior if they tell a compelling story that (1) provides hope and a roadmap to a sustainable future, and (2) helps visitors understand how their personal actions can make a difference." We need to go beyond seeing the public simply as individual consumers of knowledge. A civic engagement strategy [*Nisbet*, 2010] views individuals as potential active learners, decision makers, and participants in social and civic issues. To achieve this potential, visitors need inspiration, motivation, and empowerment, as well as knowledge.

## 9.2. THE POTENTIAL OF INFORMAL SCIENCE EDUCATION

Informal science education institutions (ISEIs) can help bridge the gap between climate scientists and the public, promoting effective public discourse about important environmental issues. In the United States, there are more than 1,500 ISEIs (science centers, museums, aquariums, zoos, nature centers, national parks, and such) visited annually by 61 percent of the population [*Inverness Research Associates*, 1996; *National Science Board*, 2012]. Furthermore, research has demonstrated that ISEIs can have a positive impact on science learning for individuals from groups who are historically underrepresented in science [*Bell et al.*, 2009].

Given that Americans spend only about 5% of their lifetime learning effort in formal, school-centered education, and only a small part of that time is focused on science learning [*Falk and Dierking*, 2010], ISEIs will continue to play a critical role in shaping public understanding of environmental science issues in the years ahead. Recent research indicates that climate change is the environmental issue of most concern to the public and that the public expects and trusts aquariums, zoos, and museums to communicate solutions to environmental and ocean issues and to advance ocean conservation [*Fraser and Sickler*, 2009; *The Ocean Project*, 2009; *Miller*, 2010; *Luebke et al.*, 2012].

Live-animal institutions, such as zoos and aquariums, attract large numbers of people of all ages (more than 175 million in the United States; www.aza.org/visitor-demographics), have strong connections to the natural world [*Belden et al.*, 1999; *The Ocean Project*, 2009], and engage many visitors who may not come with a primary interest in or educational background in science [*National Science Board*, 2012]. These centers have a unique advantage in educating visitors by making strong connections between animals and conservation issues that evoke affective responses [*Luebke et al.*, 2012].

We know from cognitive and social science research that learning is integrated with emotion [*Falk and Sheppard*, 2006] and social development [*Marcus et al.*,

2000; Brader 2006. Experiential learning in ISEIs can activate these affective connections [*Maibach*, 2012] through personal, engaging, and immersive experiences. By tapping into our emotions, these experiences activate the power of our "intuitive" learning processes, creating a more powerful impact than can be achieved by activating "analytical" learning alone [*Kahneman*, 2011]. However, although ISEI interpreters are one of the public's most trusted sources of reliable science information [*Fraser and Sickler*, 2009], research indicates that interpreters resist engaging in these discussions because of lack of confidence in their knowledge and the emotional effort required to interpret distressing topics. Interpreters need specialized training and ongoing support to help them understand climate change, its connections to the ocean, and how to relate it to the living exhibits they interpret.

## 9.3. DEVELOPING A NATIONAL STRATEGY

Ultimately, we need to take a strategic approach to the way climate change is communicated in ISEIs, with grounding in accurate climate science, a cognitive and social science evidence base, and best practices among informal science education (ISE) practitioners. Since 2007, the New England Aquarium (NEAq) has led a collaboration with an array of environmental science institutions across the United States to create a national effort to increase the capacity of ISEIs to effectively communicate about the impacts of climate change and ocean acidification on marine ecosystems. NEAq is now leading an NSF-funded partnership, the National Network for Ocean and Climate Change Interpretation (NNOCCI). This is a collaborative effort among three key areas of expertise [as recommended by *Pidgeon and Fischhoff*, 2011]:

• *Informal Science Education Practitioners*: NEAq, the Association of Zoos and Aquariums, and a core group of ISE leaders in interpreting climate change—who bring a practitioner perspective and can exponentially facilitate a learning process for additional interpreters—to develop and implement interpretation and training.

• *Climate Scientists*: Woods Hole Oceanographic Institution scientists and advisors with expertise on climate change and the ocean—who can summarize and explain what is known, characterize risks, and describe appropriate mitigation and adaptation strategies—to provide up-to-date research on forecasted global changes to the ocean, with particular attention to effects on coastal animals and habitats.

• *Learning Researchers*: Social scientists in the fields of communication and informal learning, from the FrameWorks Institute, New Knowledge Organization Ltd, Pennsylvania State University, and Ohio State University—who can bring to bear research, theory, and best practices from cognitive, communication, knowledge acquisition, and social learning theory—to conduct and disseminate research on communicating climate change to the public.

NNOCCI's design is based on best practices in informal science learning, climate literacy, cognitive/social psychology, interpretation, community building, and diffusion of innovation:

• *Interpreters as Communication Strategists*. Interpreters can serve not merely as educators disseminating climate change information but as "communication strategists" engaging in conversations with visitors based on audience research, role playing, and reflective feedback on their practice [*Nisbet*, 2010]. They can be leaders in influencing public perceptions, given their high level of commitment, knowledge, public trust, social networks, and visitor contact. On the sociological scale, interpreters can be viewed as a "tiny public" who can advocate for social change in all their interactions in other social circles [*Fine and Harrington*, 2004].

• *Communities of Practice*. The community-building and dissemination strategy for this initiative is built on considerable practical experience with climate change education in informal settings, and on the theoretical perspective of communities of practice [*Lave and Wenger*, 1991; *Wenger*, 1998]. The community of practice approach posits that learning is not an abstract or isolated activity. Instead, it is a social activity that is created through orientation, participation, exploration, reflection, and engagement in a community context. Communities of practice can enable educators to access, share, and create knowledge, and to build professional identity, relationships, and collaboration [*US Department of Education*, 2011]. The Community of Practice model will be applied throughout the "learning life cycle" of the participants, from the regional workshops, to study circles, to ongoing and online support.

• *Social Networks and the Diffusion of Innovation*. In the spread of innovation, the influence of peer networks is of primary importance [*Rogers*, 1995; *Adler and Kwon*, 2002. As a result, it is not necessary to reach everyone to effect change. "Early adopters" of a new technology or ways of thinking and behaving (introduced by innovators) influence others by creating a tipping point for change. The early majority then follow to achieve a critical mass of implementation. Social networks are important because they provide social capital where members influence and support those most closely connected to them and influence others via bridges to other networks. NNOCCI will achieve diffusion and build social capital by (1) providing shorter regional workshops, reducing the "entry cost" and providing an opportunity to try out a new approach while learning firsthand from respected peers; and using study circles to engage a critical mass of interpreters who can spread the word to others in their immediate environment, professional organizations, and in their personal social networks.

## 9.4. THE NATIONAL NETWORK FOR OCEAN AND CLIMATE CHANGE INTERPRETATION

Over the next five years, NNOCCI will build on the knowledge, success, and momentum they have generated, ultimately engaging millions of visitors in learning about climate change and the ocean. Figure 9.1 illustrates the relationships between NNOCCI's activities, the audiences they will reach, and the impacts they will achieve. They anticipate the following outcomes:

1. A new "culture of communication" about complex science, environmental and policy topics will form within the ISE community, starting with a durable, evidence-based *core story* and supporting training materials about climate change and its impacts on the oceans. A "core story" is a coherent narrative that combines an interrelated set of values, metaphors, and principles that broaden the public conversation and are consistent with the essential elements of the expert consensus. Visitor research and message testing will inform a robust interpretive strategy embodied in videos, training materials, including a self-guided on-line "e-Workshop" (now available in draft form at frameworksinstitute.org/workshops/climatechange/), and professional development activities.

2. A national network of *interpretive leaders* throughout the ISE field will be skilled and confident in communicating about climate change, and prepared to address other complex issues. Fifteen regional leaders will recruit participants for workshops and co-facilitate Study Circles. Interpretive leaders will train others, including staff, volunteer, and youth interpreters. For the growing network of participants, they will facilitate ongoing dialogue and establish an online community (see www.climateinterpreter.org). Outcomes for interpreters include increased climate literacy; skills in applying strategies, tools, and materials for educating about climate change; understanding of how to participate in a community of practice; and increased engagement in lifelong learning about science.

3. Engagement of a critical mass of *ISE institutions* with a broad national reach. Over five years, they will use study circles (see www.neaq.org/NNOCCI for details) to reach 150 ISEIs in the United States, including every major aquarium, zoos with an ocean exhibit, coastal national marine sanctuaries and national parks, every coastal state and all major urban markets (see Figure 9.2).

4. Increased *public awareness* of climate change as a salient, meaningful, and actionable topic. *Salient* refers to helping people see that an issue is important. For issues of public concern such as climate change, it means getting people to see the interdependence of humans and animals and their shared reliance on habitats. *Meaningful* denotes effectively translating concepts that enable people to make connections they did not make previously, such as appreciating fundamental mechanisms and seeing chains of cause and effect. *Actionable* means conveying who is responsible, suggesting appropriate solutions, and giving people a sense of agency in approaching problems of this magnitude. Visitors exposed to this kind of science interpretation will show increased awareness, knowledge, and engagement; the communications research we conduct ensures that the stories they tell have this effect.

5. Increased capacity of the *next generation of ocean scientists* as effective communicators. We will offer workshops for scientists at professional conferences (such as meetings of The Oceanography Society, American Geophysical Union, and American Society of Limnology and Oceanography) focusing on strategic framing and communication. Impacts for scientists will include increased awareness of the importance of effective science communication and increased skills in translating science for the public.

### 9.4.1. Target Audience

Aquarium professional and volunteer ("paraprofessional") aquarium interpreters and educators make up the first-stage audience for this initiative. This group represents participating institutions and members of an expanding network of aquariums and ISEIs interested in bringing messages about climate change and the ocean to their visitors. Young scientists are a second key audience serving as partners in communicating complex science topics. A third professional subgroup of the target audience is made up of teen interns and volunteers at participating ISEIs, which intentionally reach out to underserved populations to support science literacy and provide career development opportunities. These youth interpreters represent a high level of ethnic, cultural, and economic diversity. Teens can serve as "communication vectors" through their interactions with visitors as well as via their own social networks. They are at an important stage in their formative development of academic and career interests [*Youniss and Yates*, 2007], and form an important bridge to creating the next generation of well-informed citizens.

The public audience comprises the millions of visitors reached by participating ISEIs each year. These audiences will engage in programs, exhibits, or presentations developed by participants in the study circles and their trainees. A second public audience consists of social contacts encountered by the interpreters during their workday, leisure time, and family commitments—where

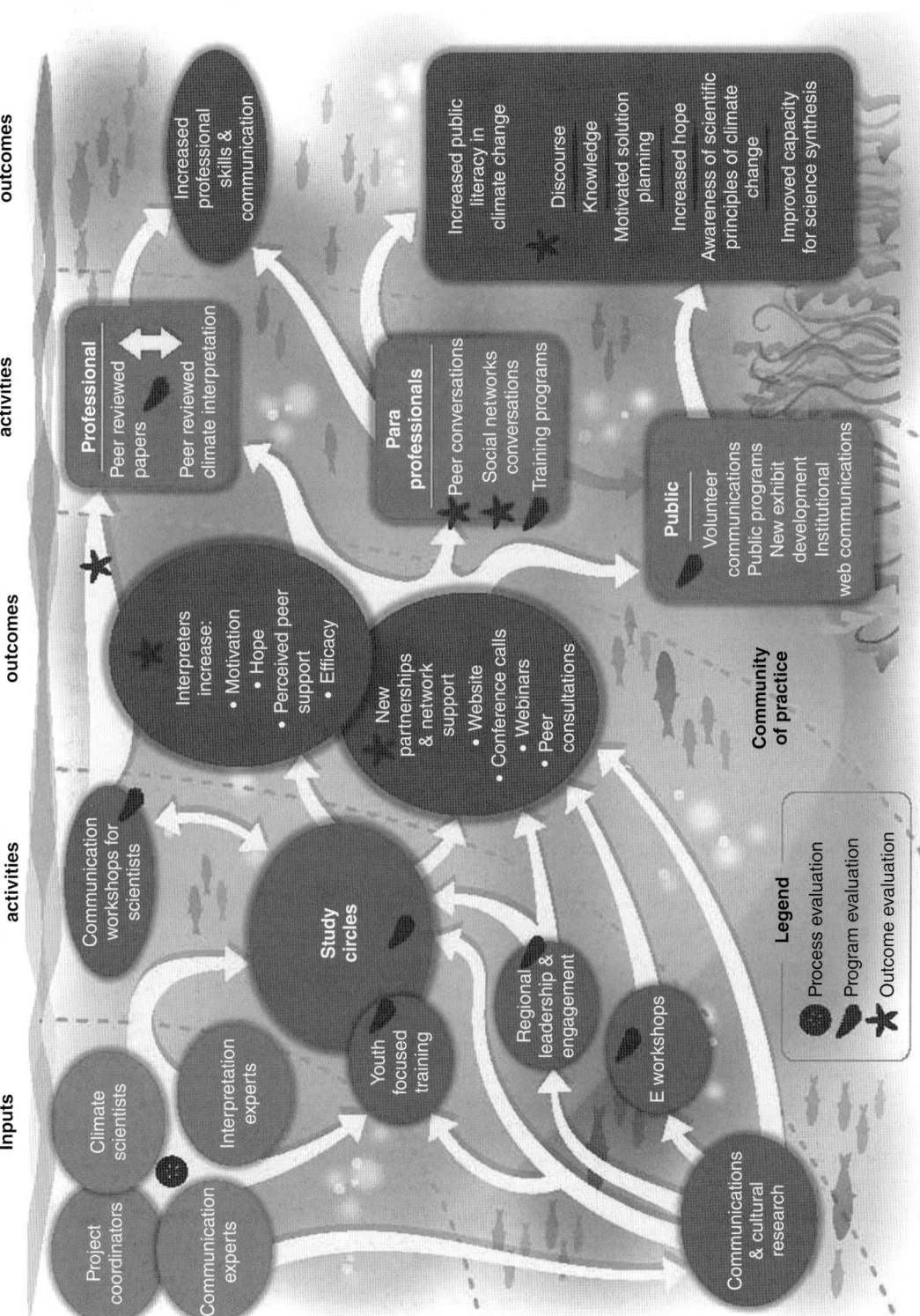

**Figure 9.1** NNOCCI logic model, showing the relationships among project inputs, activities and outcomes. For color detail, please see color plate section.

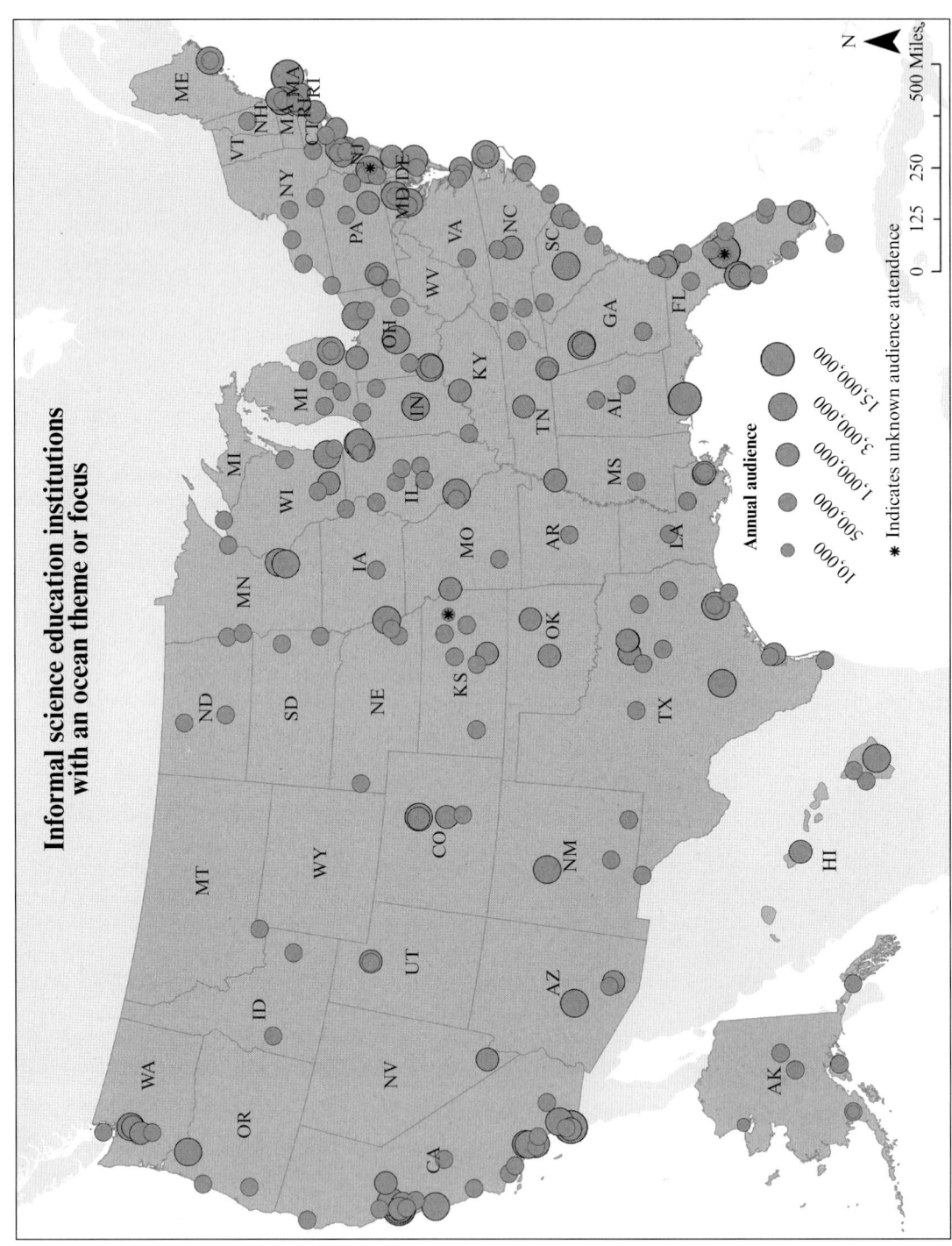

**Figure 9.2** NNOCCI has identified 212 ISEIs with an ocean theme or focus as potential participants; the goal is to reach 150 of these. For color detail, please see color plate section.

their influence and knowledge is respected and valued. A pioneering study undertaken by project evaluators [*Fraser et al.*, 2009] demonstrated that volunteerism has impacts on society far beyond the visiting population, and includes role modeling with family and peers, mentoring, behavior in other organizations beyond the volunteer site, and even social activism. This is particularly true for those who find social support within a network that validates the beliefs and values of the participant. *Fraser* [2009] demonstrated that this social validation has direct impacts across the life course and serves as an essential social support that encourages promotion of ideas and values and delivery of information related to environmental issues well beyond the walls of an informal learning institution. NNOCCI evaluators from NKO and Penn State have replicated this finding and validated methods for assessing the impact of study circles on increasing self-efficacy and social engagement in discussions about ocean climate science.

## 9.5. RESULTS TO DATE

For the initial pilot study circles, FrameWorks developed a curriculum tailored to NNOCCI, incorporating a mix of direct instruction in two key domains: (1) Strategic Frame Analysis, an evidence-based approach to communications drawing on both longstanding and experimental methods from the cognitive and social sciences, and (2) Findings and Recommendations from FrameWorks' research on how the public thinks about oceans and climate change. The curriculum included review of and guidance on materials development and practice and customized technical assistance over time. Based on other study circles, these pilots were designed to help participants incorporate FrameWorks' communications approach to effectively translate climate and ocean science. Materials were based on Frameworks' prior research on climate change and oceans, including 28 focus groups, 2 experimental surveys, 3 media content analyses, 80 cultural model interviews, and metaphor testing with more than 400 Americans.

Study circles have approximately 20 participants representing 10 institutions from across the country, along with a minimum of two climate scientists (graduate student or postdoctoral level) and two communication experts. Participants invest a substantial commitment of time (approximately 100 hours over six months) including: participation in three face-to-face workshops, readings and documentation of their attention to issues in the news, incorporation of new strategies or recommendations into ongoing interpretation and communication practice, and collaboration with fellow participants to review and critique materials and strategies. In addition to face-to-face workshops and webinars, participants receive coaching and technical assistance via Internet and telephone.

The front-end evaluation study of participants before the study circles found that they were ready and eager to learn about climate change communication and interested in building capacity within their institutions. At the same time, they were concerned about how to address the topic without discouraging or antagonizing visitors, uncertain about what level of detail to give, and less likely to engage visitors with the topic because of its complexity and a perceived threat of being challenged by those who do not accept the validity of the science. Overall, participants were seeking a consensus on the most important messages to communicate to visitors and collaboration with others to gain insight and ideas.

Impact assessments have validated the effectiveness of study circles. Most participants cited increased knowledge or understanding of cognitive and communication science or strategic framework analysis strategies and techniques. Participants gained confidence, self-efficacy, and a sense of hope in their ability to effectively communicate to the public about climate change. Most participants said they planned to share strategic framework analysis at their organizations, seeing themselves training others. Most agreed that their participation will help validate, within their institutions, the importance of climate change and the need to allocate more resources. A number of participants were applying what they learned within their "sphere of influence" at their institution, including exhibit and program development. Participants were more likely to engage others in their social network in conversations about climate change, to report increased positive feedback from these discussions, and to self-describe as confident interpreters. Three-quarters of the interpreters and all of their social network members (friends, colleagues, trainees) reported perceiving that social network members were receptive to messages that the interpreters gave, and the majority felt that these conversations left them feeling more optimistic and hopeful.

## 9.6. LEGACY AND SUSTAINABILITY

NNOCCI seeks to achieve a systemic impact across the ISE community over the next five years, creating a critical mass of participation, engagement, and impact. NNOCCI also will seek to embed its work within multiple ongoing regional and national climate change education networks. NNOCCI's legacy will include the following:

1. An evidence-based core story and supporting training materials will be incorporated in an e-Workshop, which will be widely disseminated via AZA, other professional networks, and climateinterpreter.org.

2. A national network of interpretive leaders will continue to convene and collaborate, starting with regional leaders, as part of NNOCCI's ongoing participation in the national AZA community.

3. The online community at climateinterpreter.org will continue to serve the 150 ISEIs that NNOCCI reaches over the course of the project— a critical mass with a broad national reach— and help to support further dissemination through the ISE community of aquariums, zoos, science centers, and national parks and marine sanctuaries.

4. Ongoing public opinion research will document the lasting impact of this project through increased public awareness of climate change as salient, meaningful, and actionable.

5. Young scientists from Woods Hole Oceanographic Institute and other ocean science graduate programs will bring new perspective and communication skills, enabling them to broaden the impact of their research as the next generation of ocean scientists.

Ultimately, informal science interpreters are envisioned as "vectors" for effective science communication, ocean and climate scientists with enhanced communication skills, and increased public demand for explanation and dialogue about global issues. The NNOCCI project can serve as a model not only for communicating about climate change, but for how ISEIs can address other complex environmental, scientific, and policy topics as well.

## 9.7. ACKNOWLEDGMENTS

This material is based on work supported by the National Science Foundation under Grant No. DUE-1239775. Any opinions, findings, and conclusions or recommendations expressed in this material are those of the author and do not necessarily reflect the views of the National Science Foundation.

## REFERENCES

Adler, P. S., and S. W. Kwon (2002), Social capital: Prospects for a new concept, *Acad. Manage., Acad. Manage. Rev.*, 27: 17–40.

Association of Zoos and Aquariums (2013), *Who Is Visiting the Zoo?* retrieved November 25, 2013, from www.aza.org/vistior-demographics.

Attari, S. Z., M. L. DeKay, C. I. Davidson, and W. Bruine de Bruin (2010), Perceptions of energy consumption and savings, *Proc. Natl. Acad. Sci., USA*, 107, 16054–16059.

Bales, S. N, and F. D. Gilliam (Eds.) (2004), *Communications for Social Good, in Practice Matters: The Improving Philanthropy Project*, The Foundation Center, retrieved November 14, 2013, from www.fdncenter.org/for_grantmakers/practice_matters/.

Belden, Russonello, and Stewart (1999), *Communicating about Oceans: Results of a National Survey*, retrieved November 14, 2013, from http://www.brspoll.com/uploads/files/Oceans%20summary.pdf.

Bell, P., B. Lewenstein, A. W. Shouse, and M. A. Feder (Eds.) (2009), *Learning Science in Informal Environments: People, Places, and Pursuits, Committee on Learning Science in Informal Environments*, A report of the National Research Council of the National Academies, The National Academies Press, Washington, D.C.

Brader, T. (2006), *Campaigning for Hearts and Minds: How Emotional Appeals in Political Ads Work*, University of Chicago Press, Chicago.

Brechin, S. R. (2003), Comparative public opinion and knowledge on global climate change and the Kyoto protocol: The US versus the world?, *Int. J. Sociol. & Soc. Pol.*, 23, 106–134.

Corbett J., and J. Durfee (2004), Testing public (Un)certainty of science: Media representations of global warming. *Sci. Comm.*, (26)2, 129–151.

Falk, J., and L. Dierking (2010), The 95 Percent Solution: School is not where most Americans learn most of their science. *Am. Sci.*, 98, 485–493.

Falk, J., and B. Sheppard (2006), *Thriving in the Knowledge Age*, AltaMira Press: Lanham, MD.

Fine, G. A., and B. Harrington (2004), Tiny publics: Small groups and civil society. *Sociol. Theory*, 22 (3), 341–356.

Fischhoff, B. (2007), Non-persuasive communication about matters of the greatest urgency: Climate change, *Environ. Sci. Techn.*, 41, 7204–7208.

FrameWorks Insitute (2010), *Climate Change and Oceans*, retrieved November 25, 2013, from www.frameworksinsitute.org/oceansclimate.html.

Fraser, J. (2009), An Examination of Environmental Collective Identity Development across Three Life-Stages: The Contribution of Social Public Experiences at Zoos (Doctoral dissertation, Antioch University New England, 2009), Dissertation Abstracts International, retrieved November 25, 2013, from https://etd.ohiolink.edu/.

Fraser, J., and J. Sickler (2009), *Why Zoos and Aquariums Matter: Handbook of Research Key Findings and Results from National Audience Surveys*, Association of Zoos and Aquariums, Silver Spring, MD.

Fraser, J., S. Clayton, J. Sickler, and A. Taylor (2009), Belonging at the zoo: retired volunteers, conservation activism and collective identity, *Aging and Society*, 29(3), 351–368.

Gifford, R. (2011), The dragons of inaction: Psychological barriers that limit climate change mitigation and adaptation. *Am. Psychologist*, 66(4), 290–302.

Inverness Research Associates (1996). An Invisible Infrastructure: Institutions of Informal Science Education, retrieved November 14, 2013, from www.astc.org/resource/education/infrastructure_sum.htm.

Kahneman, D. (2011), *Thinking Fast and Slow*, Farrar, Straus, and Giroux, New York.

Karl, T., J. M. Melillo, and T. C. Peterson (Eds.) (2009), *Global Climate Change Impacts in the United States: A State of Knowledge Report from the U.S. Global Change Research Program*, Cambridge University Press, New York.

Kempton, W., J. S. Boster, and J. A. Hartley (1997), *Environmental Values in American Culture*, MIT Press, Cambridge, MA.

Lave, J., and E. Wenger (1991), *Situated Learning: Legitimate Peripheral Participation*, Cambridge University Press, Cambridge, United Kingdom.

Leiserowitz, A., E. Maibach, and C. Roser-Renouf (2008), Global Warming's "Six Americas": An Audience Segmentation. Yale Project on Climate Change Communication, retrieved November 14, 2013, from http://www.climatechangecommunication.org/images/files/Six_Americas_Screening_Tool_Manual_July2011.pdf.

Lorenzoni, I., A. Leiserowitz, M. De Franca Doria, W. Poortinga, and N. F. Pidgeon (2006), Cross-national comparisons of image associations with "Global Warming" and "Climate Change" among laypeople in the United States of America and Great Britain, *J. Risk Res.*, 9(3), 265–281.

Luebke, J. F., S. Clayton, C. D. Saunders, J. Matiasek, L.-A. D. Kelly, and A. Grajal (2012), Global climate change as seen by zoo and aquarium visitors, presentation at the Chicago Zoological Society, Brookfield, IL, retrieved November 23, 2013, from http://clizen.org/survey.html.

Maibach, E. (2012), Communicating Climate Risk. From the Climate Access roundtable "Understanding and Addressing the Unprecedented Risk of Climate Change," retrieved November 14, 2013, from http://www.climateaccess.org/resource/tip-sheet-edward-maibach-communicating-climate-risk/.

Marcus, G., W. Neuman, and M. MacKuen (2000), *Affective Intelligence and Political Judgment*, University of Chicago Press, Chicago.

Miller, J. D. (2010), Civic scientific literacy: The role of the media in the electronic era, in Kennedy, D., and Overholser, G., (Eds.), *Science and the Media*, (pp. 44–63), American Academy of Arts and Sciences, Cambridge, MA.

National Oceanic and Atmospheric Administration (NOAA), (2009), *Climate Literacy: The Essential Principles of Climate Science*, retrieved November 14, 2013, from http://cpo.noaa.gov/OutreachandEducation/ClimateLiteracy.aspx.

National Oceanic and Atmospheric Administration (NOAA) (2011), *National Ocean Service: Ocean Facts*, retrieved November 14, 2013, from www.oceanservice.noaa.gov/facts/population.html.

National Science Board (2012), *Science and Engineering Indicators*, retrieved November 14, 2013, from www.nsf.gov/statistics/seind12/.

Nisbet, M. C. (2010), Civic Education about Climate Change: Opinion-Leaders, Communication Infrastructure, and Participatory Culture. White Paper presented at the Climate Change Education Roundtable, National Academies, Washington, DC, October 21–22.

The Ocean Project (2009), *America and the Ocean: Public Opinion Research of Awareness, Attitudes and Behaviors Concerning the Ocean, Environment and Climate Change*, retrieved November 25, 2013, from www.theoceanproject.org/communication-resources/market-research/reports.

Pidgeon, N., and B. Fischhoff (2011), The role of social and decision sciences in communicating uncertain climate risks. *Nature Climate Change*, 1, 35–41.

Rogers, E. (1995), *Diffusion of Innovations*, Simon and Schuster, New York.

Shore, B. (1996), *Culture in Mind: Cognition, Culture and the Problem of Meaning*, Oxford University Press, Oxford.

Tannen, D. (1999), *The Argument Culture: Stopping America's War of Words*, Ballantine Books, New York.

Trautman, C. (2007), Global warming at European museums, *Informal Learning Review*, 87, retrieved November 14, 2013, from http://www.sciencenter.org/whatsnew/d/2007-11-01%20global%20warming.pdf.

US Department of Education (2011), Connect and Inspire: Online Communities of Practice in Education, Draft Report, retrieved November 25, 2013, from http://lincs.ed.gov/professional-development/resource-collections/profile-706.

Weber, E.V., and P. C. Stern (2011), Public understanding of climate change in the United States, *Am. Psychologist*, 66(4), 315–328.

Wenger, E. (1998), *Communities of Practice: Learning, Meaning, and Identity*, Cambridge University Press, Cambridge, United Kingdom.

Youniss, J., and M. Yates (2007), *Community Service and Social Responsibility in Youth*, University of Chicago Press, Chicago.

# 10

## Opportunities for Communicating Ocean Acidification to Visitors at Informal Science Education Institutions

### Douglas Meyer[1] and Bill Mott[2]

Public opinion research by The Ocean Project has helped illuminate the opportunity that exists for scientists to partner with informal science education institutions, such as museums, aquariums, and zoos, to engage the public on issues of environmental conservation. This research, including a recent set of studies centered on the issue of ocean acidification, confirms that the public sees informal science education institutions as a trusted source for conservation information, and adds insights about public expectations during visits.

The Ocean Project has a long track record of working in partnership with aquariums, zoos, and other informal science education institutions, providing its network of more than 1,800 partner institutions with opinion research and related resources that help them communicate with their audiences on a wide range of environmental issues. In 2012, The Ocean Project conducted a set of studies related to the issue of ocean acidification, with an emphasis on identifying ways to inspire the US public to take action on the issue [*The Ocean Project*, 2012].

Ocean acidification has emerged as an issue of serious concern to aquariums, zoos, and all other institutions with a vested interest in the future of the ocean and its animals. As the world's oceans absorb increasing amounts of atmospheric carbon dioxide, pH is decreasing and acidity increasing, threatening the fundamental chemical balance of ocean and coastal waters. The National Oceanic and Atmospheric Administration (NOAA) of the US Department of Commerce refers to ocean acidification as, "an urgent environmental and economic issue," whereas former NOAA director, Jane Lubchenco, was even more colorful in her characterization of ocean acidification as climate change's "equally evil twin" [*Associated Press*, 2012].

### 10.1. METHODOLOGY

The Ocean Project's set of studies related to the issue of ocean acidification was comprised of a representative national public opinion study, a survey of professionals working within its network of partner institutions, as well as a survey of more than 3,500 members of the public as they were visiting 12 partner museums, aquariums, and zoos. These studies were completed in the fall of 2012.

The national public opinion study was completed for The Ocean Project by the globally respected market research organization, IMPACTS, whose sophisticated research integrates modeling and simulation techniques to provide highly actionable and predictive intelligence based on a sample that is fully reflective of the demographics of the US population. The surveys were primarily conducted online, but phone and intercept (face-to-face) interviews were used to verify the findings. The findings were structured using a scalar variable, representing degrees of agreement with certain statements (e.g., "On a scale of 1 to 100 with "1" being completely disagree and "100" being completely agree). The responses were aggregated and divided by the number of respondents to obtain the scalar mean. In the spring and summer of 2012, IMPACTS asked a series of questions specific to the issue of ocean acidification as part of its ongoing and overarching look at US opinions on ocean issues. Additional information about IMPACTS' methodologies can be found at http://theoceanproject.org/communication-resources/market-research/faq/.

---

[1] Principal, Bernuth & Williamson, Washington, D.C.
[2] Director, The Ocean Project, Providence, Rhode Island

*Future Earth—Advancing Civic Understanding of the Anthropocene, Geophysical Monograph 203*, First Edition.
Edited by Diana Dalbotten, Gillian Roehrig, and Patrick Hamilton.
© 2014 American Geophysical Union. Published 2014 by John Wiley & Sons, Inc.

The partner survey was developed and distributed by staff of The Ocean Project. This "opt-in" survey received more than 200 responses representing a wide array of institutions from the organization's global network. The 30-question survey focused on whether and how these institutions were addressing the issue of ocean acidification. These answers were then analyzed in combination with information obtained from in-depth interviews with representatives from 11 partner institutions, as well as a comprehensive review of the online presence of more than 50 partner institutions.

The visitor survey also was developed by The Ocean Project. For the purpose of testing the difference in response between members of the public on-site and off-site, the short survey included some of the questions from the national IMPACTS survey. The survey was loaded onto 12 iPads, which were distributed three or four at a time to nine aquariums and three science museums, where in turn those institutions collected 3,465 responses from a random sample of their visitors.[1]

This chapter will focus on key findings from this set of studies as a basis for discussing how market research can be used to shape conservation discourse in the public sphere. It will address the key findings of these studies and their implications for communicating ocean acidification, as well as consider how to apply the lessons learned from studies such as this for communicating other conservation issues.

## 10.2. KEY FINDINGS

1. The broader public is largely unaware of ocean acidification; yet, when provided with basic information about the issue they quickly became concerned.

This finding has its foundation in the national survey results, as supported by the on-site visitor intercepts. The national survey confirmed our sense that unaided awareness of ocean acidification is extremely low. On the aforementioned agreement scale (which runs from 1 [total disagreement] to 100 [total agreement]), the unaided score for "I have heard of the issue of ocean acidification" was a 14 for the public as a whole, rising only slightly to 19 when looking at recent visitors to a zoo, aquarium, or museum, and to 31 when looking only at those who already claimed concern about climate change. Yet, once provided with a simple description of the issue, concern spiked notably. With recent visitors to museums, aquariums, and zoos, for example, the agreement score on "worried about ocean acidification" went from 26 without prompting to 60 with prompting. For those already concerned about climate change, it went from 58 or moderate agreement without prompting to 78 or strong agreement with prompting. This suggests that there is a notable opportunity right now for informal science education institutions to raise awareness, and in so doing, spark appropriate levels of concern about the serious situation facing our ocean.

2. Aquariums and science museum visits often serve as activation points, offering an excellent opportunity not only for raising awareness but for inspiring action.

With the initial finding in mind, we asked 12 of our participating partner institutions to conduct on-site visitor surveys. They surveyed 3,465 visitors in less than two months using a new iPad-based approach that we were piloting (see Fig. 10.1). These on-site intercepts confirmed what the national survey had suggested, which is that on-site visitors tend to express higher levels of concern for ocean issues. A typical visitor, for example, was much more likely to express agreement with the statements, "I worry about the future health of the ocean and its animals," and "I am worried about ocean acidification." Moreover, visitors when on site claimed to have higher baseline knowledge of ocean issues, including ocean acidification, and were more likely to agree with the statement, "I am informed about ocean acidification." Essentially, visitors who were on site appeared to be much more likely than members of the visiting public, when offsite, to be interested and engaged on this issue.

Perhaps most important, the data gathered in the visitor surveys offered the strongest support to date for the hypothesis that informal science education institution visitors—especially, but not limited to aquarium visitors—are not only a self-selecting group with an interest in the ocean and ocean animals, but they are also a group whose interest is activated on arrival at one of these institutions. Across all ages, visitors indicated that they expect, trust, *and* appreciate the ocean conservation information they receive from our partner informal science education institutions, seeing receipt of such information as a key part of good experience. The importance of this last finding cannot be overstated because it is the first solid data that has been seen that advancing conservation can also be good for the "gate" at informal science education institutions.

3. Many informal science education institutions are interested, but not yet active in addressing ocean acidification with visitors.

---

[1] *The participating partner institutions were the Cabrillo Marine Aquarium, the California Academy of Sciences' Steinhart Aquarium, the Exploratorium, the Florida Aquarium, MOTE Marine Laboratory and Aquarium, National Aquarium, North Carolina Aquarium at Fort Fisher, Oregon Coast Aquarium, Pacific Science Center, Science Museum of Minnesota, Seattle Aquarium, and the Virginia Aquarium and Marine Science Center.*

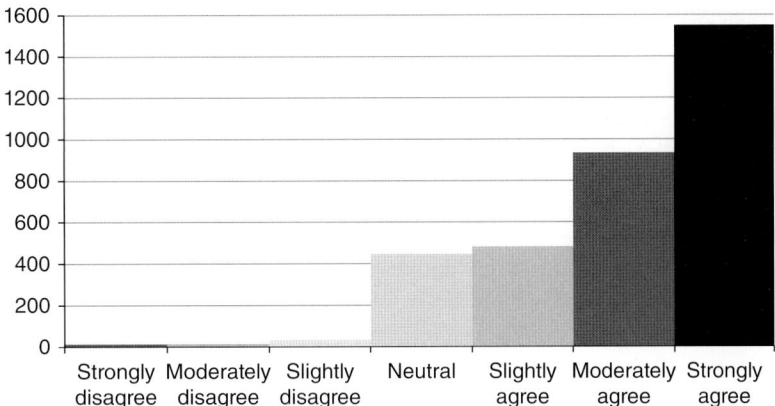

**Figure 10.1** Only 1.5 percent of aquarium and science museum visitors disagreed with the statement, *"Learning how to help conserve the ocean and its animals makes this a better place to visit,"* which a clear indication of the public's interest in learning how to be a part of conservation solutions. For color detail, please see color plate section.

We base this third and final finding on the results from the survey sent to the entirety of the partner network, as well as the series of in-depth interviews completed with representatives from a dozen leading institutions, and a comprehensive review of the online presence of 52 partner institutions.

The partner survey confirmed that the members in the network are indeed active in communicating a wide range of conservation issues with a wide range of audiences. Yet, at the same time, it also informed us that they are not yet focused on the issue of ocean acidification. Approximately 35 percent of those staff who are active in communications indicated that their institutions are already addressing ocean acidification in some way. Another 52 percent of these respondents noted that their institutions either are planning to do so, or "thinking about it."

The information gathered in the other sections of the survey, combined with the insights gained specifically from the in-depth interviews, helped to further flesh out this finding, letting us know that among those already addressing ocean acidification, only in a few cases is it currently a focal issue. More commonly, it is mentioned in passing as part of a larger conversation about climate change or in the context of threats to, for example, coral reefs. The online review provided further support because only 12 of 52 institutional websites reviewed mentioned ocean acidification, with only four of those sites offering any significant detail about the issue.

To help them in their efforts, the partners emphasized in both the survey and the interviews that they would welcome three specific types of assistance: strategy sessions to help shape their efforts, monitoring and evaluation to help measure their successes, and case studies through which to learn about the work of others.

## 10.3. IMPLICATIONS FOR COMMUNICATING OCEAN ACIDIFICATION

So how can we use some of these key findings in the efforts to develop communication about ocean acidification, especially in ways that address informal science education institution visitor interest in personal actions that contribute to a larger solution? Research suggests that not only is now the time to help put this issue on the public radar, but also that there is a remarkable opportunity for informal science education institutions to frame this issue accurately, raise awareness, and most notably, inspire action.

The public's unfamiliarity with the issue can be seen as an opportunity rather than an obstacle. Influencing public concern about well-established issues can be difficult and resource intensive, whereas emergent issues offer greater opportunities [*Druckman and Bolson*, 2011]. Climate change has been framed by the public discourse as a political issue divided along party lines (i.e., climate change is a Democrat issue), but ocean acidification is still unchartered territory. There is an opportunity to frame the public discourse in an emerging critical environmental issue.

As for how to maximize this opportunity, this recent research suggests that informal science education institutions can go further faster by aligning with public interests and expectations as they relate specifically to visits.

1. Emphasize solution steps that the public can take on their own.

Although increasing scientific literacy about the problems facing the planet is undeniably important, the research suggests that the public is more likely to be interested in the ways in which they can help solve those problems. During a visit to a museum, aquarium, or zoo, the public's interest in conservation increases, and visitors expect, trust, and appreciate suggestions from these institutions as to how they, through personal action steps, can help protect and conserve the ocean. But they show much lower levels of interest in learning about the problems. Therefore, it seems that for informal science education institutions the approach of starting with the solution, rather than the problem, is the best way to begin to address this and other conservation issues with the visiting public. In other words, the visiting public is not looking for *why* they should help as much as *how* they can help.

2. Separate "ocean acidification" from "climate change."

Even though, as noted previously, NOAA's Dr. Lubchenco famously said that ocean acidification is the equally evil twin of climate change, research suggests de-linking the two issues in public discourse. Climate change is seen as a politicized issue, and most of the US public has already decided where they stand on it. As such, there is little leeway in shifting that perception. In contrast, ocean acidification is an emergent, as yet largely undefined issue that seems to immediately spark concern. The implication is that an informal science education institution may be able to encourage similar, if not the same, beneficial behaviors by discussing them as solutions to ocean acidification, rather than as solutions to climate change.

3. Link ocean acidification to the fate of specific animals or places.

The more concrete the context, the easier it will be for the audience to understand the importance of the issue as well as its impact. Many conservation issues such as climate change and ocean health can seem abstract and have little relevance to daily life and consequently viewed as little affected by personal action. On the other hand, saving charismatic animals such as the giant panda or the American bald eagle have shown much more significant success. We can take a lesson from such past campaigns and recognize the importance of linking conservation issues to specific animal, people, or issue that informal science education institution audiences care about. For example, the issue of ocean acidification and its impact on oyster farming in the Pacific Northwest—a regionally important industry—has already caught the attention of the local population.

4. Recognize the role of informal science education institutions as trusted messengers.

Communication theories are clear on the importance of a trusted messenger—the more credible your messenger, the greater the positive impact of your message. Informal science education institutions reach a broad spectrum of people from all walks of life and they have the added benefit of being more highly trusted by the US public than almost any other conservation-related organization [*The Ocean Project*, 2009; A*ssociation of Zoos and Aquariums*, 2009]. This is not to suggest that informal science education institutions are the only trusted source, but with an annual audience of more than 200 million visitors [*Association of Zoos and Aquariums*, 2009], and the average American visiting an informal science education institution every 19 months, they are a good ally in the effort to communicate for conservation action around the issue of ocean acidification.

## 10.4. CONCLUSION

Recent research tells us that zoos, aquariums, museums, and other informal science education institutions want to do more to address conservation issues, and that the visiting public would welcome that information, especially as it pertains to ways in which they can help be a part of the solution. It seems that there is now an excellent opportunity to work with informal science education institutions to advance public awareness and action on this important issue. A next step is to develop, test, and measure the effectiveness of different approaches.

## REFERENCES

Associated Press (2012), "Ocean Acidification Is Climate Change's 'Equally Evil Twin,' NOAA Chief Says," retrieved December 12, 2013, from http://www.huffingtonpost.com/2012/07/09/ocean-acidification-reefs-climate-change_n_1658081.html,

Association of Zoos and Aquariums (2009), Visitor Demographics, retrieved March 20, 2013, from http://www.aza.org/visitor-demographics/.

Druckman, J. N., and T. Bolsen (2011), Framing, motivated reasoning, and opinions about emergent technologies, *J. Comm.*, 61(4), 659–688.

The Ocean Project (2009), America, the Ocean, and Climate Change: Key Findings, retrieved March 20, 2013, from http://theoceanproject.org/resources.php.

The Ocean Project (2012), America and the Ocean: Ocean Acidification, retrieved March 20, 2013, from http://theoceanproject.org/wp-content/uploads /2012/09/Special_Report_Summer_2012_Public_Awareness_of_Ocean_Acidification.pdf.

# 11

## City-Wide Collaborations for Urban Climate Education

Steven Snyder[1], Rita Mukherjee Hoffstadt[2], Lauren B. Allen[3], Kevin Crowley[4], Daniel A. Bader[5], and Radley M. Horton[6]

Although cities cover only 2 percent of the Earth's surface, more than 50 percent of the world's people live in urban environments, collectively consuming 75 percent of the Earth's resources. Because of their population densities, reliance on infrastructure, and role as centers of industry, cities will be greatly impacted by, and will play a large role in, the reduction or exacerbation of climate change. However, although urban dwellers are becoming more aware of the need to reduce their carbon usage and to implement adaptation strategies, education efforts on these strategies have not been comprehensive. To meet the needs of an informed and engaged urban population, a more systemic, multiplatform and coordinated approach is necessary.

The Climate and Urban Systems Partnership (CUSP) is designed to explore and address this challenge. Spanning four cities—Philadelphia, New York, Pittsburgh, and Washington, DC—the project is a partnership between the Franklin Institute, the Columbia University Center for Climate Systems Research, the University of Pittsburgh Learning Research and Development Center, Carnegie Museum of Natural History, New York Hall of Science, and the Marian Koshland Science Museum of the National Academy of Sciences. The partnership is developing a comprehensive, interdisciplinary network to educate urban residents about climate science and the urban impacts of climate change.

## 11.1. CITIES AND CLIMATE CHANGE

As urban centers continue to expand, the infrastructures on which these populations rely, including energy, water, transportation, and public health, face unique vulnerabilities to climate change. High population density, interdependent networks of infrastructure and resources, and roles as centers of industry heighten the importance of urban vulnerabilities. Furthermore cities, as centers of economic activity and dense populations, may be responsible for between 40 and 80 percent [*Satterthwaite*, 2008, references therein] of greenhouse gas emissions. Cities can simultaneously be paradigms for low-carbon living; for example, transportation sector emissions are low when population density is high, and urban planning and public transportation are robust. Cities can also be paradigms for transformative and integrated adaptation planning efforts that simultaneously improve quality of life. Responding to climate change in cities, through both adaptation and mitigation, can reduce and prevent future impacts locally and globally.

---

[1] Executive Director, Reuben H. Fleet Science Center, San Diego, California

[2] Vice President, Education and Visitor Experience, San Antonio Children's Museum, San Antonio, Texas

[3] Graduate Student, Learning Research & Development Center, University of Pittsburgh, Pittsburgh, Pennsylvania

[4] Professor, Learning Research & Development Center, University of Pittsburgh, Pittsburgh, Pennsylvania

[5] Research Analyst, Center for Climate Systems Research, Columbia University, New York

[6] Associate Research Scientist, Center for Climate Systems Research, Columbia University, New York

---

*Future Earth—Advancing Civic Understanding of the Anthropocene, Geophysical Monograph 203*, First Edition.
Edited by Diana Dalbotten, Gillian Roehrig, and Patrick Hamilton.
© 2014 American Geophysical Union. Published 2014 by John Wiley & Sons, Inc.

Cities are likely to be greatly impacted by a changing climate. Higher temperatures and an increased frequency and intensity of extreme heat events, projected for most cities globally, are likely to increase heat-related illness and mortality [*Li et al.*, 2013]. Those at greatest risk include the elderly, young children, and people with preexisting medical conditions [*ClimAID*, 2011]. A warmer climate also has the potential to impact the critical infrastructure of cities. More frequent high temperature days may cause buckling and deterioration of materials used in railway tracks and road surfaces, for example. With warmer temperatures, there will be an increased strain placed on energy systems as a result of increased demand from greater air conditioning use. Material breakdown and increased strain on the grid can both cause service disruptions in critical transportation and energy systems [*New York City Panel on Climate Change (NPCC)*, 2010]. Heat-related climate impacts in cities are exacerbated by the urban heat island effect, a condition that results in urban centers and cities being several degrees warmer than their surrounding areas [*Blake et al.*, 2011].

Cities are also particularly vulnerable to extreme, short duration rainfall events, which are projected to increase in the urban Northeast and many other regions in the future [*Horton et al.*, 2011; *Intergovernmental Panel on Climate Change (IPCC)*, 2012]. The high percentage of impervious surface cover in urban areas causes less water to be absorbed at the surface, resulting in more runoff. The increased runoff leads to flooding of streets, sewers, underground transit tunnels, homes, and businesses. More frequent heavy rainfall may cause an increase in combined sewer overflows, which can pollute urban waterways [*Horton et al.*, 2010]. Changing frequency and intensity of droughts locally and regionally also has the potential to impact the availability of water to cities.

Many of the world's largest cities are located in close proximity to the coast, making them extremely vulnerable to sea-level and coastal flooding. Coastal flooding will occur more frequently with sea-level rise alone, even without changes in storm frequency and intensity. The greatest impact from rising sea levels will be the potential for inundation of coastal areas surrounding urban centers. Similar to the effect of increased heavy rainfall, sea-level rise will cause flooding of streets, sewers, homes, and businesses. Many cities have transportation corridors that are along the coast, and these transportation systems often have underground tunnels that are vulnerable to sea-level rise. Saltwater can damage equipment at energy-generating facilities and wastewater treatment plants, which often are within the flood plain.

Finally, the characteristics of dense, urban environments further enhance the vulnerability of cities to the adverse effects of climate change. For example, the high population density of cities could amplify the spread of vector-borne disease, one potential impact of a warmer and wetter climate. In addition, cities often have large groups of residents who, based on socioeconomic factors, are poorly prepared to respond to a changing climate and plan for future impacts. The close interconnectivity of city systems also increases the risks. Failure of critical infrastructure in one system can have cascading effects on other systems (i.e., electricity outages stopping transit service.)

Given the potential impacts of global climate change on the systems on which urban populations rely and the potential for these populations to positively impact mitigation efforts, it is clear how important public engagement with these issues is for the successful development of adaptation and mitigation strategies. Sound understandings of the relevant climate science, tailored to the circumstances of particular cities and the concerns of specific communities, are fundamental to making these connections meaningful and useful in personal and civic decision making. Although the notion of global climate change strikes many non-experts as distant, abstract, or even fanciful, cities make the global local by translating distant abstractions into real consequences for land, air, water, the built environment, and public health. As city residents become more informed about how changes will directly impact them and the familiar urban systems on which they depend, they can begin to take informed action to respond to climate change. Many city leaders are wisely pursuing proactive adaptation and mitigation policies, seeing win-win opportunities to improve urban quality of life through economic benefits and enhanced infrastructure [*Rosenzweig et al.*, 2010]. These efforts will succeed only if urban residents deepen their climate literacy and meaningfully engage in issues around climate change.

## 11.2. CLIMATE CHANGE EDUCATION AT THE CITY SCALE

How then should we design educational systems to engage urban audiences in ways that activate, educate, and mobilize with respect to climate change? The conventional approach might be to teach people knowledge about climate change and then expect that they would change their behavior based on a rational understanding of atmospheric science and the impact of human behavior.

However, recent studies suggest that understanding the science behind climate change does not automatically, or even predictably, lead to individual choices that are scientifically informed or environmentally sensitive [e.g., *Shepherd and Kay*, 2011; *Kahan et al.*, 2012]. In some cases, a focus on knowledge without attention to the moral, ethical, and emotional aspects intrinsic to climate change can even result in individuals making choices that are the opposite of what a climate educator might intend [*Roeser*, 2012]. For knowledge to lead to action, climate change education may have to be broadened to address a range of factors beyond scientific information, including self-efficacy belief, perceived personal relevance, concern, and ethical or emotional responses to climate change [*Patchen*, 2010; *Roeser*, 2012; *Akerlof et al.*, 2013].

Recent educational studies also suggest that climate education may have to broaden beyond individual messaging to explicitly address how individuals see themselves as part of larger groups. Identity, social norms, and community influence are consistently predictive variables when examining environmentally friendly behaviors—often much better predictors of behavior than knowledge [*Barr*, 2007; *Nigbur et al.*, 2010; *Crowell and Schunn*, 2013]. For example, *Kahan and colleagues* [2012] found that among those who were highly knowledgeable about climate change science, political ideology was the strongest predictor of level of concern for climate change. This pattern appeared to be the result of the powerful influence of individuals' identity groups: the communities to which they are most dependent for social and physical resources.

Because people's identities and communities have a stronger effect on their behaviors than their scientific knowledge, the research from multiple fields supports a learning model that leverages community-level learning, rather than individual-level learning [e.g., *Devine-Wright, et al.*, 2004; *Shandas and Messer*, 2008]. Knowledge about climate and response to climate change should be embedded in contexts that are a part of a citizen's daily life and community in the city. There should be multiple opportunities to encounter, discuss, and share climate messaging within established social and community groups. Urban residents who are already participants in their own communities (of interest or of geographic proximity) should get the sense that "people like me" are talking about, thinking about, caring about, and responding to climate information because it is something that people like me do in this city.

As an example of how one might design for this kind of impact, consider a project where, over the course of 12 years, the city of Portland, Oregon, engaged citizens in watershed management projects by partnering with community organizations and providing expertise and leadership from both the city's public works department and the local university's urban planning faculty. The university partners, who advised the community projects, identified three main questions that drove the success of the Community Watershed Stewardship Program: "How can citizens become more involved? What is the optimal mix of local technical expertise and community capacity? And what innovations and accommodations must public agencies make?" [*Shandas and Messer*, 2008, pp. 414–415. They found that citizens' involvement in watershed management programming increased when they were included as stakeholders with real ownership from the beginning of the projects, and that "community members became more aware of the connection between their actions and the health of the environment" by working with others to improve the condition of their regional waterways (p. 414). The authors also emphasize the need for tangible results in these projects—increasing awareness and an enhanced network of community partners were important aspects of the projects' success, but real results helped to give participants a sense that their work contributed to real improvement.

## 11.3. CLIMATE AND URBAN SYSTEMS PARTNERSHIP

When the current state of climate change education across Philadelphia, New York, Pittsburgh, and the District of Columbia is reviewed, it is clear that city-wide audiences are not yet engaged in ways that will lead to effective city-scale response to climate change. However, at the same time, significant resources and efforts are currently being employed with the goal of improving the state of understanding about climate and climate change. The problem is that many of these efforts are not coordinated into a larger city-scale learning system [*Abbasi*, 2006]. Each educational effort most often moves forward with its own set of goals and approaches to learning and climate change. Although individual programs may be strong and well designed, the collection of efforts too often appears as a random assortment of projects moving in different directions and even at times contradicting each other. The result, at best, provides little synergy, and at worst, could be counterproductive. If two members of the same family were to encounter different programs using different approaches to climate change education, it is not difficult to imagine that the resulting discrepancy in understanding could lead to the promulgation of the notion that climate change itself is a debated topic.

The diversity of efforts and organizations should be turned to an advantage. Imagine if organizations across a city could be organized both in goals, message, and language. Individual projects "tuned" to each other and strategically deployed could produce resonance across the multiorganizational field of urban communities. Reinforcing experiences across multiple platforms would provide opportunities for coordinated interventions. If those interventions could also be concentrated physically or temporally, urban community members could encounter multiple reinforcing messages across several experience platforms. Such coordination would also increase the efficiency of up-scaling efforts aimed at engaging with—and voicing citizen perspectives to— urban policy makers and decision makers.

CUSP is working toward this goal by developing a collaboration of Urban Learning Networks (ULNs) designed to implement a coordinated set of integrated climate science learning initiatives across a broad spectrum of learning environments in each partner city. The ULNs bring together the wealth of organizations currently working in each city to mount a unified effort to improve city-wide climate literacy. To begin, an urban resident is thought of who throughout his or her daily activities would encounter multiple experiences that impact learning and understanding of climate science. Although not everyone will have the same experiences, given the considerable variation in daily routines, such common everyday experiences have the potential to constructively increase climate literacy through repeated exposure and common themes. Therefore, to be most effective, programming targeted at increasing climate science literacy in the general public will harness the reinforcing effects of multiple encounters across these platforms by developing a targeted and coordinated approach.

The ULNs will implement programs and curricula in such a way that they are targeted to specific community organizations and neighborhood-level groups, coordinated by presenting consistent and clear opportunities to engage with the science of climate change and concentrated across time and space. Further, the end goal of the coordinated programming (be it messaging, literacy, or engagement) will be customized as appropriate to the communities served. With such an approach, information about climate change and its potential impacts will be available through a broad range of learning experiences, providing multiple reinforcing opportunities to engage in quality climate science learning across each city. The development of a collaborative network of agencies and organizations committed to developing thematically and temporally coordinated climate science education programs that serve their interest or geographically based members or audiences will result in consistent learning experiences about climate science to the diverse publics of the cities. The result will be a relevant, city-wide approach to improving the state of climate literacy in the urban environment.

That said, the complexity of climate science and the current minimal state of public knowledge presents an extremely large hurdle when approaching the development of climate change education programs. With the resources assembled, what then should be the approach to climate science learning? Based on recent research, it is clear that a purely knowledge-based individual target approach will be insufficient to meet the learning needs of the intended audience [*Shepherd and Kay*, 2011; *Kahan et al.*, 2012]. Rather, to be most effective, a set of design principles will be needed that consider not only cognitive learning but also the social and emotional aspects of climate change and rely on the power of community-level learning [*Patchen*, 2010; *Roeser*, 2012; *Akerlof et al.*, 2013]. We are working with three core design principles, rooted in the learning sciences and educational research literatures:

1. **Framing for Relevance**. Not all audiences have the same values or come into a conversation from the same point of view. For more conservative audiences, environmentally friendly behavior is much more attractive when it is framed as an issue of economic or energy security, such as seeking independence from foreign oil, or an act of patriotism, such as buying goods made in the United States. For audiences that experience oppression, framing climate change as an issue of justice can be a good way to tap into what people are already passionate about and personally affected by, because economically disadvantaged communities are often more heavily affected by extreme weather events. It is this personal relevance and connection to personal passions that serves as the starting point for programming.

2. **Participation**. Participation is one of the most powerful mechanisms for learning and is one of the primary forms of engagement in informal learning. Participation refers to hands-on, interactive, and authentic (i.e., in-context) experiences that lead to learning, development of attitudes, and the making of personal connections. Urban residents may participate in a hands-on learning activity that educates about the dynamics of heavy rain events and combined sewer overflows, or they may work together to care for an urban garden and learn about changes in plant growth seasons and ranges.

3. **Systems Thinking**. Fundamentally, climate change will be experienced through its impacts on the urban systems on which citizens depend. Understanding this, however, will require a level of system thinking that is not taught in schools, nor is it generally recognized as a strong attribute in most highly educated adults. To understand systems, people need to be able to engage with the intersections of science, society, individual passions, and unfamiliar topics. For example, heat wave mortality is partly a climate science issue, partly an air quality issue (because in the urban Northeast air quality is often poor during heat events), partly an infrastructure issue (because electrical systems may be more prone to failure precisely when air conditioning is most needed), and partly a social issue (because vulnerability to heat depends on a number of societal factors). Climate change is thus a "socio-scientific" problem, meaning that it is more than simply a "problem of science" [*Houser*, 2009].

These three design principles have been used to design the overall approach for project programming. CUSP programming begins with framing issues in a way that is relevant personal passions. What do city residents care about? What are the issues, topics, and activities with which residents personally and socially identify? Framing educative efforts for relevance to the concerns urban residents have that intersect with climate change provides the necessary starting point with which to begin a deeper conversation about climate change and city systems.

Establishing and promoting programming as a conversation is a key element of participation. Learning is more than just receiving a message; it means engaging with others who are more, less, or equally expert on the various intersecting ideas. Facilitating space for those conversations to happen and generating a shared language and trust among the people having the conversation,= will be a, important aspect of CUSP.

The next step in that conversation is to connect personal interests to the urban systems on which they rely. In an urban setting, few issues, activities, or hobbies are not dependent on one or more citywide systems. Whether one identifies as a gardener, advocates for public transportation, or worries about the cost of electricity, there is always an underlying urban system. By connecting urban residents' passions to this urban system, CUSP programming will take the first step in developing urban residents' understanding of how climate change can disrupt the systems that connect them to the things they care about. Frames of relevance and processes of participation in the multi-issue field of climate change adaptation and mitigation should encourage understanding of systems, just as systems thinking should support the development and evolution of relevant frames and processes of participation. To understand climate change as it impacts the systems that people depend on, people need the opportunity to engage with the intersections of science, society, individual passions, and unfamiliar topics as they relate to those systems. The CUSP model and principles are coordinated to provide multiple and varied opportunities for this type of engagement.

This model of connecting personal passions, urban systems, and a changing climate through participatory learning experiences forms the basis of the CUSP programming approach. However, a final element is needed. When faced with the issues associated with climate change, people often have mixed reactions. On the one hand the sheer scope of the issues can result in a feeling of powerlessness. The negative emotional connotation of this reaction does not bode well for the chances of either continued engagement with learning about climate science or meaningful engagement with further climate issues. On the other hand, there are those who respond with the need to engage in immediate individual action. Although this result may be no worse than neutral, the reality of the limited impact of individual action can lead once again to a sense of powerlessness. This is particularly true in the context of the city. The networked organizations that make up the heart of CUSP ULNs are uniquely situated to connect urban residents to the governmental, organizational, and community-based efforts that enact adaptation and mitigation strategies on a community-wide level. Examples of engagement around coastal flood risk in New York City at each of these three levels include the Mayor's Office of Long Term Planning and Sustainability, The Trust for Public Land, and local environmental justice groups. By interfacing with these and other groups, urban residents who want to take action can find a meaningful outlet and those who might feel powerless in the face of the issues are provided with the support network they need.

## 11.4. CURRENT PROGRESS AND FUTURE WORK OF CUSP

CUSP calls for a targeted, coordinated approach that relies on connecting personal passions to urban systems and how they will be impacted by a changing climate. Because this requires engaging multiple organizations and efforts and because these efforts need to be aligned, participants will be defined by time, space, and interest rather than a "target audience." A bottom-up approach has been taken by focusing on the general public rather than key decision makers because the long-term success of any adaptation or mitigation efforts will rely greatly on what the average person does. The project will focus on local impacts matching the need for personal engagement. Taking that into account, participants should be urban residents who have common values and shared interests who gather in a space that can be occupied by multiple organizations simultaneously. Given the large role that has been identified for community identity, the most appropriate target for the CUSP approach is a community, such as a neighborhood, a group of people who care about air quality issues, or a club of beekeepers. By engaging with the various communities in the cities, multiple existing programs can be can used to coordinate content and materials, while the shared interests and values of each community provide the opportunity to engage around shared passions, all which results in interactions between individuals and within communities that increase climate literacy and overall engagement with climate change issues.

Community itself can be broadly defined, but CUSP has identified four different community types that they will support through their programming:

• Virtual communities: the ever growing digital communities facilitated by the Internet and social media

• Temporal communities: those communities of shared values that coalesce around time-limited events such as festivals

• Physical communities: the physical geographic neighborhoods that make up the city

• Communities of practice: the shared communities of those engaged in and aligned with climate change education

As of this writing (April 2013), CUSP partners have been focused on developing a set of digital tools, festival kits, neighborhood programming, and professional workshops designed to support each of these communities. First, the digital tools are focused around the development of an online mapping system that will allow ULN members to use the interests and concerns of the digital audiences they serve to select appropriate overlays for the map, enabling users to select information that interests them. ULN members will also integrate the map into their programming, from uploading citizen science data to sharing stories and pictures via the map.

Second, the temporal communities are supported by a library of festival booth activity kits that ULN organizations will use to tap into festival-goers' curiosity. The kits offer hands-on activities related to topics such as the temperature effects of alternative roofing materials, the carbon footprint of mass-transit versus car-centered transit systems, and urban stormwater management. With kits distributed among the booths of several community organizations, visitors will have multiple opportunities for interactive learning that catalyzes conversations about climate change and their city.

Third, the neighborhood community group is establishing a pilot site in the city of Philadelphia to test collaborative programming opportunities for physical communities. Programs currently active within the pilot neighborhood (at libraries, recreation centers, railway stations, and so on) are developing a shared set of learning goals to be incorporated into the broad set of programs offered. These programs are not all necessarily climate-based educational programs. Rather they represent the mix of experience opportunities (gardening, childhood literacy, school assembly programs) currently offered. Within the pilot neighborhood, each of these educational assets will provide a coordinated set of climate learning opportunities within a three-month period.

Fourth, the communities of practice platform is developing new tools to build the capacity of ULN member organizations to effectively deliver climate change education programming to their specific memberships, target audiences, and interest- or need-based communities. A recurring theme heard from ULN members is that these groups want to improve their capacities to help their target audiences examine climate issues and make the most appropriate choices. The communities of practice group has developed and tested a training workshop for ULN community organizations and developed educational modules they can use with their audiences. The modules highlight choices and tradeoffs at a community level.

The plan is to iteratively design and document, using principles of design-based research [*Barab and Squire*, 2004], each of the climate change learning platforms, one in each of the four cities. Developers and educators are in biweekly contact with one another to ensure that each platform is being developed in a flexible enough framework to be successfully implemented in each of the other three cities after the initial round of design-based research is complete. Implementation will be conducted through design-based implementation research [*Penuel et al.*, 2011] integrating the different platforms into a single, city-scale learning environment for climate change. As the learning scientists, developers, and educators are iterating the designs of each platform; the climate scientists are developing local climate projections and working with the ULNs to identify areas of vulnerability and opportunities to adapt in each city. Finally, the evaluators are tracking progress at each step and will ultimately identify the extent to which the program has been successful in reaching audiences and changed knowledge of and response to climate at the city-scale.

CUSP aims to coordinate the efforts of multiple organizations and efforts, themed around and targeted to those issues and topics about which particular communities are passionate. These learning platforms will connect urban residents' passions to the urban systems that surround and support them and explore the climate science underlying the threat posed by a changing climate. By providing multiple opportunities to encounter reinforcing learning experiences, connecting and coordinating efforts across multiple organizations to create the opportunity for resonance, and situating these efforts in community and tapping into personal passions, CUSP will develop a unique, effective and substantial format for reaching across the urban landscape to engage city residents and their communities in climate science and the impacts of climate change.

## REFERENCES

Abbasi, D., (2006), *Americans and Climate Change: Closing the Gap Between Science and Action*, Yale School of Forestry and Environmental Studies, New Haven, CT.

Akerlof, K., E. W. Maibach, D. Fitzgerald, A. Y. Cedeno, and A. Neuman (2013), Do people "personally experience" global warming, and if so how, and does it matter? *Global Environ. Change*, 23(1), 81–91, doi: http://dx.doi.org/10.1016/j.gloenvcha2012.07.006.

Barab, S., and K. Squire, (2004), Design-based research: Putting a stake in the ground. *J. Learn Sci.*, 13(1), 1–14.

Barr, S. (2007), Factors influencing environmental attitudes and behaviors: A U.K. case study of household waste management, *Environ. Behav.*, 39(4), 435–473.

Blake, R., A. Grimm, T. Ichinose, R. Horton, S. Gaffin, S. Jiong, et al., (2011), Urban climate: Processes, trends, and projections, in *Climate Change and Cities: First Assessment Report of the Urban Climate Change Research Network*, C. Rosenzweig, W. D. Solecki, S. A. Hammer, S. Mehrotra, (Eds.), (pp. 48–

81), Cambridge University Press, Cambridge, United Kingdom.

ClimAID, (2011), Responding to Climate Change in New York State: The ClimAID Integrated Assessment for Effective Climate Change Adaptation in New York State, C. Rosenzweig, W. Solecki, A. DeGaetano, M. O'Grady, S. Hassol, and P. Grabhorn (Eds.), *Annals of the New York Academy of Sciences* 1244 (1)2–649.

Crowell, A., and C. Schunn, (2013), The context-specificity of scientifically literate action. Public Underst. Sci., DOI: 10.1177/0963662512469780, retrieved November 20, 2013, from http://www.lrdc.pitt.edu/schunn/research/papers/Crowell-Schunn-PUS2013.pdf.

Devine-Wright, P., H. Devine-Wright, and P. Fleming (2004), Situational influences upon children's beliefs about global warming and energy, *Environ. Ed. Res.*, 10, 493–506.

Horton, R. M., V. Gornitz, D. A. Bader, A. C. Ruane, R. Goldberg, and C. Rosenzweig (2011), Climate hazard assessment for stakeholder adaptation planning in New York City, *J. Appl. Meteorol. Clim.*, 50, 2247–2266.

Horton, R., C. Rosenzweig, V. Gornitz, D. Bader, and M. O'Grady (2010), Climate Risk Information, in Climate Change Adaptation in New York City: Building a Risk Management Response, C. Rosenzweig, and W. Solecki, (Eds.), (pp. 147–228), *Annals of the New York Academy of Sciences*, 1196.

Houser, N. (2009), Ecological democracy: An environmental approach to citizenship education, *Theor. Res. Soc. Ed.*, 37(2), 211–214.

Intergovernmental Panel on Climate Change (IPCC) (2012), Managing the risks of extreme events and disasters to advance climate change adaptation, in *A Special Report of Working Groups I and II of the Intergovernmental Panel on Climate Change*, C. B. Field, V. Barros, T. F. Stocker, D. Qin, D. J. Dokken, K. L. Ebi, et al., (eds.), Cambridge University Press, Cambridge, United Kingdom.

Kahan, D. M., E. Peters, M. Wittlin, P. Slovic, L. L. Ouellette, D. Braman, et al., (2012), The polarizing impact of science literacy and numeracy on perceived climate change risks, *Nature Clim. Change*, 2, 732–735.

Li, T., R. M. Horton, and P. Kinney (2013), Increasing net annual temperature-related mortality in a warming climate, *Nature Climate Change*, doi:10.1038/nclimate1902.

New York City Panel on Climate Change (NPCC), (2010), Climate change adaptation in New York City: Building a risk management response, C. Rosenzweig and W. Solecki (Eds.), prepared for use by the New York City Climate Change Adaptation Task Force, Annals of the New York Academy of Sciences, New York.

Nigbur, D., E. Lyons, and D. Uzzell (2010), Attitudes, norms, identity and environmental behaviour: Using an expanded theory of planned behaviour to predict participation in a kerbside recycling programme, *Brit. J. Soc. Psych.*, 49(2), 259–284.

Patchen, M. (2010), What shapes public reactions to climate change? Overview of research and policy implications, *Anal. Soc. Iss. Pub. Pol.*, 10(1), 47–68.

Penuel, W. R., B. J. Fishman, B. H. Cheng, and N. Sabelli (2011), Organizing research and development at the intersection of learning, implementation and design, *Educ. Res.*, 40(7), 331–337.

Roeser, S. (2012), Risk communication, public engagement, and climate change: A role for emotions, *Risk Anal.*, 32, 1033–1040.

Rosenzweig, C., W. Solecki, S. A. Hammer, and S. Mehrotra (2010), Cities lead the way in climate-change action. *Nature*, 467, 909–911, doi:10.1038/467909a.

Satterthwaite, D. (2008), Cities' contribution to global warming: Notes on the allocation of greenhouse gas emissions,. *J. Environ. Urbanization*, 20, 539–549.

Shandas, V., and W. B. Messer (2008), Fostering green communities through civic engagement, *J. Am. Planning Assoc.*, 74(4), 408–418.

Shepherd, S., and A. C. Kay, (2011), On the perpetuation of ignorance: System dependence, system justification, and the motivated avoidance of sociopolitical information, *Personality and Soc. Psych. Bull.*, 102, 264–280, doi: 10.1037/a0026272.

# 12

## On Bridging the Journalism/Science Divide

### Bud Ward*

Scientists and the journalists covering their work have had a strained relationship for as long as there have been scientists and journalists covering their work. There's nothing wrong with that, and in fact it's the way things should be and were meant to be, with each discipline bringing to the relationship the same kind of committed independence that marks the best in their respective fields. Truth is, if the public at large is to understand and support responsible science and responsible journalism, journalists need scientists, and scientists need journalists. And it's particularly true with so many important public policy issues, including, but surely not limited to, climate change, which is so heavily based on sound science.

Deep down, neither the journalist nor the scientist might enjoy acknowledging that they actually share much in common, in particular concerning the essential values that the best among them bring to their field. With their shared commitments to verification, authentication, transparency, and, if you will, "truth," and with their thirst for discovering and learning new information, there would seem to be lots of ways for the two disciplines to get along and learn from each other. But at the same time, and particularly given important changes well under way in both fields, many journalists and scientists are just beginning to get the message about the science/journalism nexus. They are doing so at a time when traditional news organizations, in particular (but not limited to) major metropolitan daily newspapers and weekly news magazines, are undergoing wrenching and in many cases foundation-shaking changes. All of which makes their task all the more important, albeit even more difficult.

The *National Association of Science Writers' (NASW)* "official" guide, *A Field Guide for Science Writers*, relates that "the cure for fear and loathing of science is neither propaganda nor persuasion but knowledge" [*Blum, et al.*, 2005, p. v]. That knowledge, Timothy Ferris writes, is "conveyed, preferably, in stories that capture and reward an audience's attention," a storytelling talent that many hard-driven scientists find completely alien to their formal educations and daily work lives.

"Writing well about science requires, first of all, bridging the jargon gulf, acting as translators between the sciencespeak of the researchers and the short attention spans of the public at large," the three NASW editors write in that book. "But great science writing doesn't stop there. You can paint an awesome picture of space exploration with all its glittering astrotoys, but you also have a responsibility to probe its failures. You can point out the benefits of genetically modified crops or the mapping of the human genome, but you also must explore their potential to do harm. It's not enough to focus on the science itself; the best reporting also discusses safeguarding the public from the risks of the new knowledge and talks about the cost of Big Science and who has to pay for it" (*Blum, et al.*, 2005, p. vii).

The challenge arises in that the revolutionary underlying foundational changes under way in the mass media over the past 15 years or so—what many experts consider just the beginning of a decades-long transition on how the public accesses and processes serious news and information—that appear to many to be wreaking havoc in particular on coverage of science and climate science issues. Journalism blogs lit up in early 2013 when *The New York Times* first announced that it was reorganizing its newsroom and doing away with the designated science and climate "desk" that specialized in coverage of those issues. Just two months later, the "old grey lady," as the

---

*Editor, The Yale Forum on Climate Change & The Media*

*Future Earth—Advancing Civic Understanding of the Anthropocene, Geophysical Monograph 203*, First Edition.
Edited by Diana Dalbotten, Gillian Roehrig, and Patrick Hamilton.
© 2014 American Geophysical Union. Published 2014 by John Wiley & Sons, Inc.

newspaper long has been called by admirers, shook the environmental journalism timbers again; it abandoned its online "green" blog that had been the venue of many freelancers whose copy, although worthwhile and closely followed by many in the environmental field, could not be published in the print edition. In both the cases, the *Times* maintained, to widespread skepticism, that its moves wouldn't come at the expense of outstanding climate and environmental coverage. Even the paper's in-house "Public Editor"— ombudsman—wasn't buying into that explanation. As still the nation's most widely respected daily newspaper and the single outlet most likely to influence and shape coverage by media elsewhere, the *Times* is not alone in paring back the staff resources and real estate it makes available for climate change science news.

Reporter *Christopher Zara* [2013] wrote a report in the *International Business Times* concerning the "particularly exigent" challenges facing science journalists in the modern media climate. In an article headlined "Remember Newspaper Science Sections? They're Almost All Gone," Zara, citing data from the *Columbia Journalism Review*, reported that the number of US newspapers with weekly science sections had fallen from 95 in 1989 to 19 currently. "That's a big drop, even for one of the fastest declining industries in the country," he wrote [*Zara*, 2013, p. 2]. He reported Bureau of Labor Statistics data of a 40 percent decline in newsroom employment over the past decade.

"Science writing is a specialty, and the more specialized the field, the more highly skilled a writer has to be," Zara reported. "It's for just that reason, insiders say, that newspaper executives sometimes come to the conclusion that their science sections should be sacrificed in lieu of more general-interest reporting" [*Zara*, 2013, p. 3].

Compounding the challenges facing sound science reporting is the reality that effective coverage also involves a facility with economics issues, with moral and ethical dimensions, and with policy and political considerations and implications. All of those, of course, can be part and parcel of good climate science journalism. That the reporters having to make sense of often complex science stories now must do so in an era of shrinking public attention spans, 140-character "Tweets," 10-second sound-bites, and 24/7 always-on social media only further complicates their challenges. Don't think so? Just try telling the story of methane "tipping points" in an 800-word news story read by many on a display no larger than their Smartphones, let alone in a Facebook post, or a 140-character tweet most likely to reach a key demographic needing to know.

In addition to competition from other news organizations and various "new" media for the public's seemingly shrinking attention span, there is also an important internal dynamic under way in virtually every major news organization, print or broadcast. It involves that competition among colleagues, mentioned previously, for the finite and dwindling "news hole" in the case, for instance, of a daily newspaper or for "air time" in the case of local television news. As mentioned, this competition for the editors' favor—with the reward, for instance, being front-page "real estate," ideally "above the fold" (at least when newsstand sales were critical) —pits science news against crime news, local sports news, city hall news, medical news, entertainment news ... the full gamut of issues a general circulation news organization might be interested in covering. The competition, needless to say, can be fierce, and science issues don't always compete well against, let's say, a local car chase or bank hold-up.

As noted, fewer newspapers still maintain a regular and dedicated science section, with fewer than two dozen continuing the practice. Science reporters remaining full-time in their newspaper newsrooms are finding their responsibilities expanding to include not only social media blogging and tweets but also an expanded range of subject areas—for instance, covering climate, medicine, space, technology, and health—at the time most of those fields are becoming more specialized.

## 12.1. HARD NEWS ... AND EXPLANATORY JOURNALISM

That straight news story about the report released earlier that morning or yesterday generally is considered a "hard news" or "day-one" story. In a number of cases, good reporters will follow-up those initial stories with subsequent ones, reporting additional details, clarifications, elaborations, and subsequent related events.

In some cases, the news "peg" for such a story might be as simple as the prestaged release of a report, a perhaps unscheduled event or, for instance, bit of testimony or a critique of a previous story. In other cases, an often equally valuable type of reporting can result in explanatory journalism. There is no particular news peg motivates the reporter or editor in this case—just recognition of a need to explain some idea or some concept that they feel needs further clarification.

Let's consider a specific example of explanatory journalism, and an outstanding example at that, one dealing with climate science. In a February 18, 2013, wire service report, Associated Press science correspondent *Seth Borenstein* [2013] raised the question of how to rationalize "scant snowfall and barren ski slopes" and "a whopper of a blizzard" ... with scientists and others pointing in both cases to climate change as "the culprit" (p. 1).

"The answer lies in atmospheric physics," Borenstein continued, explaining through science why the seemingly contrasting perspectives are not the "brazen contradiction" some portray them to be [*Borenstein*, 2013, p. 4].

"Shorter snow season, less snow overall, but the occasional knockout punch" [*Borenstein*, 2013, p. 9]. That's the quotation Borenstein attributed to Princeton University climate scientist Michael Oppenheimer, who added "That's the new world we live in."

The story didn't break news, and Borenstein or Associated Press did not seek to do so in reporting it.

Another example of explanatory journalism may be an article by *Seattle Times* veteran science reporter *Sandi Doughton* [2013]. The headline for that article: "Science students learn to tell stories."

Doughton reported on a course founded by University of Washington graduate students called "Communicating Science to the Public Effectively." A goal of the course?

"Teach young scientists how to share their passions for cosmology, chemistry, or evolutionary biology without putting people to sleep." She characterized the course—"one of several springing up across the country"—as being "fueled by a new generation of researchers who see public outreach as integral to their jobs" [*Doughton*, 2013, p. 8].

Doughton listed climate change, energy policy, resource conservation, and medical ethics as examples of "today's pressing issues" with science at their foundation. She quoted an on-campus science communicator as saying "There's a strong sense that we are not adequately preparing graduate students to face the professional world they are going to be joining… . Scientific leadership and solid communications skills are intrinsically linked." At the same time, Doughton said such courses "remain rare," and she wrote that one master student in the University of Washington course "signed up out of frustration" [*Doughton*, 2013, p. 19].

"I got tired of people's eyes rolling to the back of their head when I tried to explain what I did," student Juliana Houghton told her. She said another student studying the physics of stellar evolution found people excited to hear he is an astronomer … "but can quickly get lost in details about wavelengths, luminescence, and computer modeling" [*Doughton*, 2013, pp. 22–26].

Doughton's article also reports a University of Washington science writing teacher cautioning that "A mistake many scientists make is to view the public as empty vessels, waiting to be filled with the knowledge that will inspire them to line up behind the same agendas as scientists," even if it's only to support funding for scientific research. In reality, "people make decisions based on many factors, including emotions and values. Scientists who approach public outreach with a strident agenda can turn people off," Doughton wrote. She also reported a course student-instructor's commitment that their course "focuses more on conveying the students' excitement about science than pushing a point of view" [*Doughton*, 2013, pp. 31–34].

Hardly the stuff of hard-news reporting but no less valuable all the same. The Borenstein and Doughton examples are precisely the kind of in-depth explanatory reporting that issues such as global climate change demands. And it's the kind of somewhat nuanced climate science reporting that requires reporters and editors with a certain level of scientific understanding and know-how and also a commitment to a more informed citizenry. That is the rub: it's therefore the kind of reporting that many fear could be lost or diminished amid the current economic woes facing journalism, the often excessive pursuit of news that "entertains" more than it informs, the shrinking newsroom staffs, and the associated newsroom "brain drain."

## 12.2. TIPS FOR SCIENTISTS IN WORKING WITH MEDIA

There are no hard-and-fast rules for dealing with the media. All reporters are individuals, all of them different, even within the same news organization. The best way to work with a reporter is to work with that reporter over a period of time. Not just days or weeks, but months and years. In helping a reporter become a better reporter, a good scientist is also making himself or herself a better scientist, one more capable of communicating responsibly with an audience they might not otherwise reach. There are some basic essentials, however, that scientists might do well to consider as they engage with news media representatives and others in an effort to communicate beyond the scientist's own narrow sphere of specialists:

1. Good reporters are on the job 24/7. So from the moment they enter your office for an interview until they leave it however much later, assume that can use everything you say. Camera not yet on? Tape recorder not yet activated? It doesn't matter: Assume whatever you say, unless the reporter and you have agreed in advance of its being uttered, can be used. Terms such as *not for attribution*, *off the record*, and *background* are commonplace for many reporters tilling the political, crime, business, or other beats and, mercifully, less common in science reporting. But less common does not mean nonexistent. For those in the business of providing information to reporters—sources that is—it's important to recognize that reporters even among themselves don't universally share a common definition of those terms. It's critical that a reporter and an interviewee agree *in advance* of the words being uttered or the document shared about just what a requested ground rule is and what it means. Taking time in advance of imparting the information will save lots of time and headaches down the road.

2. Don't be afraid to ask a reporter the deadline he or she is working on. Particularly if you can't at that moment take

a reporter's phone call and questions, ask their deadline. And then honor it by getting back to them in time to help them meet their deadline. If you're not the right person for them to be speaking to on a particular subject—not your area of expertise—just say so, and point them to the proper expert if you can.

3. There's nothing wrong with your telling a reporter "I don't know." If that's your honest answer, they should honor that. Perhaps their question is not worded as artfully as it might be: Ask them if what they're really asking is ABC rather than XYZ. If so, you'll have helped *yourself* out.

4. Deadlines. They're the raison d'être in the news business. For many bloggers and for reporters constantly under the gun to "feed the beast" of social media such as Facebook or Twitter, those deadlines come constantly, incessantly, pretty much as the second hand sweeps. For a reporter working on a daily deadline for the next morning's paper, there may be a few hours to spare, and for the weekly or monthly business publication reporter, perhaps even longer.

Knowing a reporter's deadline—asking what it is and then keeping a commitment to help the reporter meet it—can be key in the scientist/reporter relationship. Just merely asking a reporter his or her deadline and promising to get back to him or her before it can be ingratiating on its own, but only with that timely follow-through. If you promise to get back to a reporter within a certain time, do you all can to keep that commitment, even it involves you providing them your cell phone or home phone number.

5. Even in today's rapidly changing media environment, many reporters come to a story with the "Five Ws and an H" in mind: Who, Why, Where, What, When, and How. Journalists sometimes feel that scientists instinctively "bury the lede" (the most important element of a news story) by putting their conclusions at the end of their research report, rather than at the beginning, as is the practice in journalism. Think about it: Asked by a family member or friend "How was your day?" are you more likely to begin telling them "We've discovered humans are eight times more susceptible to a particular toxin than are mice used in drug testing?" Or are you likely to reply by regaling them with the details of your methodology, testing procedures, and verification strategies before telling them the bottom line about humans' being more vulnerable? Reporters will want to lede (an intentional spelling media use among themselves to distinguish from "lead," just as in "hede" instead of "head" for headlines) with the bottom-line conclusion. And later on, let's hope to get to the important caveats and qualifiers, etc.

6. You're dealing with an area you've studied and researched for years. The reporter of course has far less background in your specialty area. After explaining an issue in some detail, it can be effective to politely ask, "Now, do you think you understand that point? It's a really important one in understanding this whole issue." You can even have the reporter parrot-back to you their understanding of what you've explained. In which case, don't be hesitant to reply, "Well, I think you basically have it, but there's one important nuance that I think you may have overlooked... ." Done properly and with tact, your interview with a reporter can be a valuable learning experience for both of you.

7. Don't hesitate to tell a reporter that his or her question(s) goes into a specialized area of science distinct from your own, and you're reluctant to provide an answer. Again, "I don't know" can be an entirely appropriate response.

8. A reporter might well encourage you to go beyond the scientific or technical aspects of your research and to comment on the public policy implications of it. "So does this mean you prefer a carbon tax to a 'cap-and-trade' system of regulation?" This can get dicey, but if you choose—as a citizen and *not* as a scientist—to venture into policy implications of your work, make sure the reporter knows and understands that you are removing your scientist's hat and speaking as a citizen. You're entitled to do so of course, but make sure your stipulation is clear, and that as a citizen your opinions may be no more valid—and no less—than those of any other citizen.

9. Don't expect a reporter to send you a pre-publication of pre-broadcast copy of the story they are working on. Some might be willing to do so, but many others, consistent with longstanding journalistic principles and practices, may not be. On the other hand, some reporters may welcome the opportunity to share with you in advance of publication a few sentences or even a paragraph or two to make sure they "nailed it." They're not asking you to be their editor or to change their areas of emphasis ... they're asking you to confirm that they actually captured a technical point you made. That can save both of you a lot of grief, but in some cases a reporter may choose not to go that far. It's their call in this case.

10. You know the audience(s) you most want to reach with your message about your work, and the reporter should know the demographics of the audience their coverage best reaches. When the two overlap substantially, the communications hurdles become more manageable for each of you. The language you are most comfortable with because you work with it day in and day out—International Panel on Climate Change (IPCC), oxides of nitrogen, adaptation and mitigation, climate sensitivity or climate dynamics, or even terms such as *theory*, *enhance*, and *skeptic* may mean one thing to you and something entirely different for the reporter or his or her audience. Best to know and compensate where you can,

and when in doubt, make sure you and the reporter are speaking the same language.

11. You see a glitch in the final product once it's broadcast or published? Keep your reaction in sync with the seriousness of the glitch. It may warrant your contacting the reporter to discuss, and it may also eventually warrant your going to the editor to express your concern. Tact and courtesy are important in this case too, so let your reactions be scaled to the seriousness of the mistake. Immediately going over or "around" the reporter to a higher authority can be counterproductive to your further dealings with the reporter.

12. In dealing with a reporter on virtually any issue, a news source can be on the offensive or on the defensive. If the reporter is calling you on deadline seeking your comment on your research widely criticized by an outside authority, you clearly are playing defense. And the chances of your scoring while on defense are not high. Playing offense with a reporter or news organization might involve your contacting weeks or months before your research ever sees the light of delay, backgrounding them when they are not on deadline and perhaps not even working on your story. It might involve prepping them for the time you are ready to have your research go public, so this might be an instance where close consultation with your university or employer's public affairs professionals can be helpful.

13. You are under no firm obligation to address a reporter's specific question—either in an interview or during a press conference—and not go beyond it. It's best to reply in some reasonable fashion to the question, but then go along with what you believe the real point needs to be. This is a cat-and-mouse game, and many reporters likely will come back to you if they feel you have flat-out dodged their question. That's their prerogative, but it's yours to emphasize what you truly feel is most important.

14. To use a baseball metaphor, take a lot of at-bats. Not every scientist's every encounter with a journalist (or a city council member or senator for that matter) is going to end up as a home-run. Or triple, or double, or even single. But as the late climate scientist Stephen Schneider used to emphasize, taking a lot of at-bats—and learning from each experience and from each dealing with a specific reporter—will lead to hits. And, more often than not, to a respectable batting average too.

## 12.3. RESOURCES FOR SCIENTISTS IN DEALING WITH MEDIA AND VICE VERSA

If the inherent tensions in the news source or news reporter relationship weren't enough on their own to prompt some sleepless nights on the part of all involved, the troubling aspects of the current news media "revolution" certainly compound those challenges.

So much so that those serving as providers of news and information to the news media increasingly are finding new and innovative, but sometimes troubling, ways of going around the gates and barriers frequently imposed by serious journalists and their editors (and done so in an entirely responsible way from a journalistic standpoint).

Ready access to the Internet by virtually anyone with a computer and a connection affords one way for scientists to get their messages out independently from the news media. They can do so through their own blogs or websites or, in some cases, through those of their employers. They of course can also do so—and more and more are moving in this direction, though it's by no means yet a groundswell—through social media outlets, particularly Facebook and Twitter.

Those approaches are, mind you, best seen as complements, and not as alternatives, to ongoing efforts to engage traditional news organizations when appropriate as part of a science communications effort. But unlike the days when those same media outlets were the choice for communicating with the broad public and those wanting to communicate with the public, they now are just one of a number of options, each with its own strengths and limitations.

From the basic standpoint of sound communications, targeting and prioritizing one's optimal audiences will continue to be an important first step, regardless of which specific medium or outlet(s) is then pursued to reach them. Numerous guides and resources—many of them available online and without charge—are available for using each the various "new" media outlets, and doing so need not be complicated notwithstanding some learning curve with each. Making use of the social and new media options over time will become an important component in any science communication effort. So might as well get started now.

Along with those kinds of tools for using various media options most effectively, a number of published and firsthand face-to-face resources also are available. Among the latter is one already fairly well-known in climate science circles, the Leopold Leadership Program, Stanford Woods Institute for the Environment (http://leopoldleadership.stanford.edu/), which annually awards fellowships to researchers "to help them translate their knowledge to action" by improved communications.

Along with various one-day workshops held over the years aimed at improving communications among climate scientists and reporters and editors or, separately, aimed at boosting broadcast meteorologists' understanding of climate science, a number of useful books have been written with a goal of improving science and media interactions, many of them with a specific focus on climate science.

One popular and worthwhile entry in this category is *Don't Be Such a Scientist: Talking Substance in an Age of Style*, written by one-time marine biology professor Randy Olson, now a filmmaker. A co-founder of the Shifting Baselines Ocean Media Project, designed to link scientists and Hollywood in the interest of better understanding of global oceans issues, Olson brings wit, humor, and insight to his work, as evidenced by the five chapter titles of this book:
- Don't Be So Cerebral
- Don't Be So Literal Minded
- Don't Be Such a Poor Storyteller
- Don't Be So Unlikeable
- Be the Voice of Science!

Another useful entry in this category of communications guides is *A Scientists' Guide to Talking with the Media: Practical Advice from the Union of Concerned Scientists*.

Co-authored by long-time independent broadcast science journalist Daniel Grossman and UCS media director Richard Hayes, the paperback offers scientists in sights on "why reporters do what they do," on how scientists can "master the interview," and on choosing the most appropriate communications tools.

Climate scientists in particular might find helpful the still-available "Mediarology" website (http://stephenschneider.stanford.edu/Mediarology/Mediarology.html) developed by the late Stanford University charismatic climate scientist Stephen H. Schneider, who died in 2010. That site offers a range of firsthand experiences and insights based on Schneider's personal experiences as an avid climate science communicator.

From the standpoint of media, one of the early "bibles" on science communication came in the 1973 book *Precision Journalism: A Reporter's Introduction to Social Science Methods*, written by long-time journalist and journalism educator Philip Meyer, who taught at the University of North Carolina. Meyer has been a leading proponent of how responsible media can use approaches analogous to the scientific method to improve their science coverage.

Another invaluable guide for science writers—and, for that matter, for those who want to better understand science writers and their modus operandi—is the 1989 *News & Numbers: A Guide to Reporting Statistical Claims and Controversies in Health and Related Fields*. Written by the late health and science journalist Victor Cohn, this compact and highly readable paperback offers insightful advice on how the media can (and must) do a better job communicating on issues involving, for instance, the statistics of environment and of risk and the use of statistics in reporting on politics, economics, and democracy.

Also from the perspective of the media, the National Association of Science Writers' *A Field Guide for Science Writers"* is considered an essential read for reporters tilling those fields. It's edited by a team consisting of one-time reporters and journalism educators, including a Pulitzer Prize–winning science journalist—Deborah Blum, Mary Knudson, and Robin Marantz Henig—and it has insights tailored for those working with small newspapers, large newspapers, broadcast media, trade publications, and more.

Another valuable shelf item in this category is *The New York Times Reader: Science & Technology*, written by long-time Indiana University science writing professor S. Holly Stocking and various *New York Times* reporters, including Andrew C. Revkin, William Broad, Cornelia Dean, Carl Zimmer, and Sandra Blakeslee, all respected science journalists.

In the end, some scientists can be and are excellent communicators, outstanding ambassadors to the non-scientific community—media, public officials, citizens—about the importance of their work. Not all scientists fit that mold, and those lacking the interest and desire should not be forced into a situation that would only make them and their professional colleagues uncomfortable.

What's important is that those most comfortable and most skilled at communicating with the media and with the public be encouraged to do so, and that the talent pool with those inclinations continues to expand. Those less well-suited should feel no such compunction, so long as the message about the science is getting out in a responsible way.

At the same time, keep in mind that just as nature abhors a vacuum, so too do reporters and editors abhor a news vacuum: That void will be filled one way or the other, and too often by a pro- or con- advocate far less dedicated to an evidence-based explanation than the relevant scientist might be. It's part of the rationale that is leading more and more science educators to acknowledge and heighten essential communications skills as a key component in the education and training of scientists dealing with scientific issues having great public policy salience.

Scientists, science communicators, and educators are not alone in seeking guidance on effective communications practices and strategies in the quickly changing news media landscape of the twenty-first century. Reporters and their editors also are looking for insights on how best to till these challenging fields. The author of this chapter, while he was researching and writing for this AGU monograph, was invited also to contribute a feature article in a publication of environmental journalists, addressing the question of whether journalism today is well suited to cover the climate change issue. It is reprinted here with permission.

### COVERING CLIMATE: ARE JOURNALISTS UP TO THE TASK? MORE IMPORTANTLY—IS JOURNALISM?

#### *Bud Ward*

"How well is American journalism poised to cover the climate change story now?"

The question is especially timely given expectations for final release over the coming months of federal research agencies' "National Climate Assessment," the nomination of former EPA air director Gina McCarthy as the new EPA administrator and of a new yet-to-be nominated NOAA administrator, and of the scheduled initial release this fall of IPCC's Fifth Assessment Report.

What's more, rapid and troubling changes in the journalism field continue apace: *The New York Times*' recent decision to eliminate its specialized environment and climate "desk," and the paper's scrapping of its "Green" blog, which had been an outlet for freelancers, along with the continued cuts in specialized environmental coverage at other newspapers and the demise of valued outlets such as The Phoenix in Boston.

Let's say, for purposes of argument, that there is an adequate supply of journalists and journalist-wannabes—many of them no doubt among SEJ's own members—eager to sufficiently cover the innumerable challenges that make this uniquely "generational" set of climate challenges so journalistically demanding.

Granted that the bench could always be longer, and deeper too. But the talent is there for the asking, ready now and tomorrow and in the coming decades to take on the reporting and editing challenges associated with all that comes under the infinite umbrella of human-caused climate change.

But that begs the question, which, rightly so, involves not "journalists" but rather the institution of "journalism." It's a much more thorny question, and the answers are likely to be less self-gratifying, and far less comforting.

Can an Institution in Crisis Cope with a Climate Story?

If we assume American journalism is in fact "well poised," that still leaves open only the question of just how well poised.

No doubt there could always be more—more column inches, more air minutes, more continuing education opportunities, and gobs more Twitter tweets and Facebook friends and followers and all-things-digital. They're there, as stated earlier, "for the asking." The issue comes down to whether their editors and news managers and salary payers—and their audiences—indeed are doing the asking.

But let's not get ahead of ourselves.

It's in fact cruel irony that these most pressing, yet in some ways incomprehensible, climate challenges come at a time of historic unease and transition in the media business. That's a term perhaps more apt now than ever before to the day-to-day practice of journalism in the U.S.

There are, after all, sound reasons that climate change is considered a "generational issue," and, what's more, a "wicked" one both for communicators to speak to and for the public at large to understand and confront.

No matter how effective today's journalists and today's journalism are in tackling the climate conundrum, the problems it presents may well be ultimately unfixable. We can manage them, yes. But "solve" them? Most likely no.

Climate change poses a broad set of insidious societal, economic, political, and, yes, environmental challenges that our children and theirs will still be wrestling with long after we've passed, or dropped, the baton. And wrestling, that is, as direct and indirect victims of its adverse impacts, but also as journalists and other communicators and educators striving to better inform the public generally.

With its most serious and most highly visible impacts still remote for many both in time and in distance, and with carbon dioxide, the most important single greenhouse gas, both odorless and invisible, it's hard for many to visualize the myriad problems careening down the road toward us. Further compounding the challenge is the difficulty in assessing blame, and with it responsibility—the fact that there's no single or even institutional "black hat," no Megabucks Inc., no single villain.

There's no data to suggest that the late cartoonist Walt Kelly had climate change in mind when, in a 1971 Earth Day strip, his iconic "Pogo" first penned the line, "We have met the enemy, and it is us."

But he might as well have.

That the issue transcends traditional "environmental" problems and encompasses everything ranging from public health to infrastructure, from national security to international competitiveness, and from daily weather to yearly snowmelt and sea-level rise only increases the challenges facing the media.

And those are the very challenges that the "institution" of journalism appears now resolutely unprepared to address and resolve.

What Breed of Journalists to Get the Job Done?

What skills might today's and tomorrow's reporters covering climate most need?

To do the job exceedingly well, they need to have in their DNA a lot of "ink in the veins" journalism, a sharp nose for news, and a keen knack for storytelling.

Being able to discern the real significance of the latest so-called "ground-breaking" research will be key, so they will need a lot of science smarts too and will need to know how to best vet and qualify those so-called "breakthroughs" often not deserving of that name.

That's a start, but it's far from enough. They'll need more than a tad bit of social psychology, and more than a dollop of aquatic biology to go along with a nugget of atmospheric chemistry. And of course physics. And statistics. Oh. Did we mention economics? And international diplomacy and law? And the study of cognition and of persuasion methodologies?

The qualifications list goes on. But the key point is that more so than any previous issue they have been charged with addressing, the climate change issue deeply embodies an expansive net of diverse issues and specialty fields, many of them far afield from their past experiences and the courses most of today's journalists pursued and excelled at as students.

So into this daunting challenge comes journalism, circa early 21st Century: An era of shrinking news holes, vanishing hard news outlets, paring down of science beats and desks and other specialty coverage in preference for more "GAs," and a waning of American attention spans, geared now more toward infotainment, 140-character tweets, and "e-blasts" than to heavy-lifts of daily serious reading and reflection.

You get the picture.

So it's not only fair to wonder—it's actually irresponsible to ignore—the question of whether American journalism today in too many ways drives some to leave the once-hallowed bastions of outstanding reporting to do just that—outstanding reporting.

Leave traditional journalism in order to do truly good reporting on climate change? More than a few have done just that, and their names are well recognized among environmental journalism savants. Names like Phil Shabecoff, Bill McKibben, Ross Gelbspan, Eric Pooley, Andy Revkin.

The belt-tightening, eyeballs-obsessed, "BREAKING NEWS" over-emphases characteristic of so many news organizations today imposes on quality journalism the kinds of constraints no self-respecting and enterprising journalist can conceivably welcome. Or long endure?

But while these are pitfalls primarily of the "institution" of journalism, it's critical too that journalists themselves accept some responsibilities on their own part: Too many of us, let's be honest, took news-writing and feature-writing classes, and other journalism courses knowing full well that in doing so we could avoid those danged statistics, mathematics, science, and physics electives we did our best to ignore. Those, that is, that might have given us some of the skills now so urgently needed in the media to report knowledgeably on an issue as daunting as our warming climate.

The same kinds of motives that drove those brilliant scientists to so diligently avoid journalism and philosophy and civics—their dread of words and letters and grammar and essay-writing—motivated many journalists to shun the science and math-based courses and expertise the sciences so demand.

**Democracy Demands We Close the Information Gap**

Did I mention here perhaps the most significant and troublesome shortcoming of contemporary journalism in so far as its dealings with our warming climate?

It's the issue of the steadily widening information gap, the chasm between those now and in the future having access to the world of rich and authoritative online information on climate change and its implications … and those not having, and in some cases of course not wanting, such information.

It's not traditional journalism but rather the Web and blogs and social media drill-downs and podcasts and more that serve well those already having a climate knowledge bounty, and still wanting more.

But it is within the repertoire of traditional journalism to serve also that much larger group of news—and information—havenots. Our democracy demands we do so.

Until journalists and their institutions of journalism measure up, how can anyone maintain that the democratic principles that attracted so many of us to journalism can now accommodate a response to the simple question posed other than with an abrupt "Poorly"? And then doubt that we now must go about doing all we can, individually and collectively, to rectify those shortcomings?

This column is reprinted with permission from *SEJournal*, the magazine of the Society of Environmental Journalists (http://www.sej.org), where it was initially published in the spring 2013 issue.

Notwithstanding some highly troubling cutbacks in recent years in the media resources dedicated to outstanding science journalism—fewer specialized reporters, fewer "science sections," broader areas of assignment in the face of increasingly specialized scientific pursuits—there have been some recent examples of efforts to improve the depth of coverage.

CBS News, for instance, in 2012 began a partnership with a respected environmental scientist, M. Sanjayan, bringing him into its flagship news operations to report and comment as the network's "Science and Environmental Contributor." What makes the collaboration so interesting—and so unusual in the "mainstream" journalism world—is that Sanjayan has continued in his capacity as lead scientist with The Nature Conservancy, in effect wearing "two hats" while appearing to casual television viewers as a traditional and unaligned observer. Given Sanjayan's considerable experience and skills in communicating with general audiences on complex environmental and science issues, the partnership no doubt has the potential to improve environmental and science communication for CBS News audiences. But it's precisely the kind of church-and-state collaboration—pairing a "straight news" operation with a representative of an outside policy advocacy interest organization—that traditionally has raised the hair on the backs of hard-boiled news executives. A pity, some might say, that CBS News couldn't, or didn't, find a full-time "credentialed" journalist to fill that need.

## REFERENCES

Blum, D. M. Knudson, and R. Marantz Henig (2005), *A Field Guide for Science Writers: The Official Guide of the National Association of Science Writers*, 2nd ed., Oxford University Press, New York.

Borenstein, S. (2013), "Climate contradiction: Less snow, more blizzards," retrieved December 12, 2013, from http://www.usatoday.com/story/weather/2013/02/18/climate-contradiction-less-snow-more-blizzards/1927893/.

Doughton, S. (2013), "Science students learn to tell stories," retrieved December 12, 2013, from http://seattletimes.com/html/localnews/2020628506_talkingsciencexml.html.

Ward, B. (2012), "Bridging the journalism/science divide, Professions seek new ways to collaborate," retrieved December 12, 2013, from http://www.sej.org/publications/sejournal-sufall12/bridging-journalism-science-divide.

Ward, B. (2013), "Covering climate: Are journalists up to the task? More important—Is journalism?" retrieved December 12, 2013, from http://www.sej.org/publications/sejournal-sp13/covering-climate-are-journalists-task-more-important-journalism.

Zara, C. (2013), "Remember Newspaper Science Sections? They're Almost All Gone," retrieved December 12, 2013, from http://www.ibtimes.com/remember-newspaper-science-sections-theyre-almost-all-gone-1005680.

# INDEX

Note: Italicized page locators indicate figures; tables are noted with *t*.

AAAS. *See* American Association for the Advancement of Science
Absolutist theory, 36
Academic information and counseling, lack of, for Native American students in geoscience programs and, 71
Acidification of ocean. *See* Ocean acidification
Active learning
   CYCLES and ICE-Net and, 22
   DataStreme courses and, 24
   teacher professional development and, 21
   Teachers as Watershed Researchers project and, 25
Advertising, for undergraduate research opportunity programs, 79
Advisors, REU programs and, 82
Affect
   assessment practices in geoscience education and, 48–49
   learning outcomes and, 43
Affluence, 6
Africa, AIDS in, 5
Agriculture, reimagining and reengineering, 2
AGU. *See* American Geophysical Union
AIDS, 5
Air mattress analogy, carbon dioxide accumulation and, 34
AISES. *See* American Indian Science and Engineering Society
Alaska Natives
   traditional knowledge incorporated in geoscience education and, 69
   underrepresentation of, in the geosciences, 68
Alternative theories, 36
American Association for the Advancement of Science, 48, 49*t*
American Geophysical Union, 25, 68, 74, 92

American Indian communities, teacher professional development for climate change education in, 22–24. *See also* Native Americans
American Indian Higher Education Consortium, 73–74
American Indian reservations, REU on Sustainable Land and Water Resources and community-based participatory research on, 85–86
American Indian Science and Engineering Society, 68, 79
American Indian students, GEMscholar Program for, 86–87
American Meteorological Society, mission of, 24
American Society of Limnology and Oceanography, 92
AMS. *See* American Meteorological Society
Analogies
   defined, 34
   reasoning with, 34–35, 37–38
   drawbacks of, 34–35
Anchors Project (NSTA), 10
Animal species, human activities and loss of, 2
Antarctic Peninsula, 3
Anthropocene. *See also* Community-driven research in the Anthropocene; Teacher professional development in the Anthropocene
   atmosphere and, 4–5
   civic understanding of advancing, argument for, 41
   assessment websites of relevance to, 49*t*
   humanity and, 5–6, 53
   K-12 science education framework and, 9–17
   conclusion, 16–17
   critical framework elements, 10–11
   genesis and grounding of, 9–10

   overall goal, 10
   ten recommendations for designing standards and curriculum, 11–16
   land and, 2
   ocean and, 2–4
   onset of, 1
   teaching students about, in research experiences, 83–84
   ethics training and, 83
   place-based REU, working with communities, 83–84
   three undergraduate research experiences related to, 84–87
   GEMscholar Program, 86–87
   Research Experience for Undergraduates on Sustainable Land and Water Resources, 85–86
   SOARS Program, 84–85
Aquarium professionals and volunteers, public awareness of climate change and, 92
Aquariums, Ocean Project in partnership with, 99
Arctic Ocean, demise of summer sea ice in, 3
Arctic tundra, 2
Argumentation, evidence-based, scientific reasoning and, 35–36, 38
Argument from evidence, K-12 science education framework and engaging in, 12
Assessment
   best practices, recommendations for, 49–50
   Backward Design, 49–50
   cultural validity, 50
   instrumentation and design, 50
   learners with diverse abilities, 50
   CIRES Outreach and Education Group and, 28
   for civic understanding initiatives, importance of, 42
   of cognition, 43

Assessment (cont'd)
  current practices in geoscience
      education, 47–49
    affect, 43, 48–49
    cognition, 48
  of decision making, 43
  defining, 41
  future efforts in , agenda for, 50–51
  Native-focused, for the geosciences,
      73–74
Assessment instrument types, review,
    44–47
  qualitative approaches, 44–46
    drawings, 45–46
    interviews and focus groups, 46
    observations, 46
    short response questions and
        essays, 44–45
  quantitative approaches, 46–47
    Likert-type questions, 46–47
    multiple-choice questions, 47
Assessment Tools in Informal Science, 48
Assessment websites, civic
    understanding of the
    Anthropocene and, 49t
Associated Press, 112, 113
Association of Zoos and Aquariums, 91
ATIS. See Assessment Tools in
    Informal Science
Atlantic Ocean, fishing and energy
    extraction industries in, 3
Atmosphere
  Anthropocene and, 4–5
  climatic system interconnections
      and, 33
  human release of $CO_2$, warming
      oceans and, 3–4
Atmosphere science, Literacy Principles
    developed for, 43
Attitudes, decision making and, 43
AZA, 95, 96

Backcasting, in climate change
    education, 35
Backings, arguments and, 35
Backward Design, 42, 49–50
Balloon analogy, carbon dioxide
    accumulation and, 34
Band end points, K-12 science
    education, 11
*Benchmarks for Science Literacy*
    (AAAS), 10
Big Science, writing about, 111
Biomes, of North America, 2
Biosphere, climatic system
    interconnections and, 33
Blackfeet community, Cultural Validity
    in Geoscience Assessment
    project and, 73–74

Blackfeet Community College, 73
Blacks, underrepresentation of, in the
    geosciences, 68
Blakeslee, Sandra, 116
"Blanket analogy," as alternative
    explanation for greenhouse
    effect, 34
Blogs, 115, 119
Blum, Deborah, 116
Borenstein, Seth, 112, 113
Broad, William, 116
Buhr, Susan, 26

Cabrillo Marine Aquarium, 100
Carbon, Earth's energy balance and, 4
Carbon dioxide
  air mattress and balloon analogies
      and accumulation of, 34
  atmospheric
    human-induced increase in, 4–5
    ocean acidification and, 99
    warming oceans and human releases
        of, 3–4
Carbonic acid, ocean acidification
    and, 3–4
Career choice issues, for Native
    American students in
    geoscience programs, 71
Carnegie Museum of Natural History,
    103
Case studies, Earth Science-related,
    K-12 science education and, 15
CBS News, 119
Cellular telephone subscriptions,
    worldwide, 6
Center for Hazards Assessment,
    Response and Technology, 85
Center for Multiscale Modeling of
    Atmospheric Processes,
    84, 85
Chaos theory, K-12 science education
    framework and, 15
CHART. See Center for Hazards
    Assessment, Response and
    Technology
Circle of learning principle, Geoscience
    Alliance and, 68
CIRES. See Cooperative Institute for
    Research in Environmental
    Science
CIRES Outreach and Education Group
  choosing roles you will enjoy, 28
  climate and oceans teacher
      professional development
    extreme ultraviolet variability
      experiment geomagnetism in
      MESA classroom, 27
    scientific inquiry on Tibetan
      Plateau, 27

    waterspotters, 27
  connecting geoscientists and
      educators, 26–28
  developing two-way relationships, 28
  finding partners who use best
      practices, 28
  including assessment and
      evaluation, 28
  meeting needs of schools, 28
  purposes of, 26
  thinking of outreach partner as you
      do your other collaborators, 28
Cities, 2
  climate change and, 103–104
  climate change education and,
      104–105
  coastal, 2–3
Citizen-science, 57
Civic engagement strategy, understanding
    climate change and, 90
Civic understanding of planet Earth
  assessment best practices,
      recommendations, 49–50
  assessment instrument types, review
      of, 44–47
    qualitative approaches, 44–46
      drawings, 45–46
      interviews with focus groups, 46
      observations, 46
      short response questions and
          essays, 44–45
    quantitative approaches, 46–47
      Likert-type questions, 46–47
      multiple-choice questions, 47
  cognitive, affective, and behavioral
      considerations, 42–43
    learning outcomes, 42–43
      affect, 43
      cognition, 43
      decision making, 43
  current assessment practices in
      geoscience education, 47–49
    affect, 48–49
    cognition, 48
  evaluation and assessment of
    defining, 41
    importance of, for civic
        understanding initiatives, 42
  future assessment efforts, agenda
      for, 50–51
Claims, arguments and, 35
Climate, urban systems partnership
    and, 105–107
Climate and Urban Systems
    Partnership, 103
  current progress and future work
      of, 107–108
    bottom-up approach, 107
    community types in, 107–108

goal of, 103
Urban Learning Networks and, 105–106
  framing for relevance, 106
  participation, 106
  systems thinking, 106–107
Climate change. *See also* Global climate change; Public dialogue on climate change
  cities and, 103–104
  cognitive and social psychology of, 89–90
  community-driven approaches to, 57
  excessive atmospheric $CO_2$ and, 4
  geoscience national conferences and theme of, 68
  human citizenry and adapting to, 5
  indigenous communities and, 61
    impacts, experiences and actions, 62–64
  "indigenuity" and adapting to, 63
  land loss in the Louisiana bayou and, 60
  ocean acidification and, 99, 102
  polarized nature of belief in, 54
  "Six Americas" survey and teachers' attitudes toward, 23
  as socio-scientific problem, 106
  traditional knowledge systems coupled with science concepts and coping with, 69
  wild rice, White Earth Nation and, 60
Climate change education. *See also* Urban climate education
  in American Indian Communities, teacher professional development for, 22–24
  argumentation promoted in, 35
  at the city scale, 104–105
Climate change story, journalism and coverage of, 117–118
Climate education, scientific reasoning and, importance of, 32–33
Climate literacy, 89
  avoiding misconceptions and, 33
  city-wide, 106
  enhancing, 31
    through application of scientific reasoning, 33–37
      epistemological reasoning, 36–37
      evidence-based argumentation, 35–36
      reasoning with analogies, 34–35
*Climate Literacy: The Essential Principles of Climate Science* (NOAA), 23
Climate science
  complexity of, 33, 37
  Literacy Principles developed for, 43

CMMAP. *See* Center for Multi-scale Modeling of Atmospheric Processes
Coal, carbon in, 4
Coastal flooding, cities and, 103
Coastal populations, 2–3
Co-creation pathway, wild rice, White Earth Nation and, 60
Cognition
  assessment practices in geoscience education and, 48
  learning outcomes and, 43
Cognitive psychology, of climate change, 89–90
Cognitive science, assessment instruments and, 51
Coherence
  CYCLES and ICE-Net and, 22
  DataStreme courses and, 24
  teacher professional development and, 21
  Teachers as Watershed Researchers project and, 26
Cohn, Victor, 116
Co-learning, 60, 61, 69
Collaborations
  assessment and, 51
  Earth Science-related, K-12 science education and, 15, 16–17
Collaborative curriculum, K-12 science education framework and, 16–17
Collaborative processes, devaluing of, science-society gap and, 54
Collaborative research, team-structured, REU programs and, 82
Collective participation
  CYCLES and ICE-Net and, 22
  DataStreme courses and, 24
  degree of, teacher professional development and, 21
College degrees, global population with, 6
Colonial history, ecology interwoven with, 62
Colorado State University, 84
*Columbia Journalism Review*, 112
Columbia River Inter-Tribal Fish Commission, intergovernmental collaborations and, 63
Columbia University Center for Climate Systems Research, 103
Communalism, 54
Communication norms, science-society gap and, 54
Communities of practice, CUSP and, 107, 108

Community-based research projects, related to Anthropogenic issues, working with communities on, 83–84
Community colleges, recruiting from, for REU programs, 79, 81
Community-driven research in the Anthropocene, 53–64
  closing science-society gap and, 55–57
  common elements of community-driven science, 57–61
  epilogue, 61
  participatory approaches and, 61
  paths for connecting science and society, 56
  science-society gap and, 54–55
Community-driven science, 61
  advantages of, 57
  common elements of, 57–59
    beginning with a community-question, 57–58
    co-learning, 59
    embracing multiple priorities, 58
    leveraging diversity, 59
    successful participatory projects, 58–59
    valuing community knowledge, 58
  examples, 59–61
    land loss in the Louisiana bayou, 60
    meningitis in the Sahel, 59–60
    wild rice and White Earth, 60–61
Community influence, climate change education and, 105
Community knowledge, valuing, 58
Community of Practice model, 91
Community-question, beginning with, 57–58
Community Watershed Stewardship Program (Oregon), 105
Complexity, K-12 science education framework and, 15–16
Concept inventories, 48
Concept mapping, 23, 57
Conservation issues, informal science education institutions and, 100, *101*, 102
Content focus
  CYCLES and ICE-Net and, 22
  teacher professional development and, 20–21
  Teachers as Watershed Researchers project and, 25
Controversial topics, addressing in teaching about the Anthropocene, 20
Cooperative Institute for Research in Environmental Science, 84
Corals, ocean acidification and, 4

Core Ideas, K-12 science education framework
  case studies in, 15
  designing standards and curriculum, recommendations, 12
  presentation of, 11
  progressions of teaching and learning, 16
Core stories, defined, 92
Corn, 2
Council for Undergraduate Research, 77
Counterarguments, 36
"Covering Climate: Are Journalists Up to the Task? More Importantly–Is Journalism?" (Ward), 117–118
Creativity, global digital technology and, 6
Croplands, 2
Crosscutting concepts
  defined, 19
  K-12 science education framework and
    case studies in, 15
    designing standards and curriculum, recommendations, 12
    presentation of, 10–11
    progressions of teaching and learning, 16
Crosscutting themes
  Earth and Space Science standards and, 19–20
  U.S. science standards and, 43
Crow Tribe (Montana), NOAA data, mitigation of waterborne microbial health risks and, 63
Cryosphere climatic system interconnections and, 33
Cultural competence, 69
Cultural humility, 69
Cultural identity issues, obstacles in geoscience education for Native Americans and, 70
Cultural validity
  assessment best practices and, 50
  defined, 73
Cultural Validity in Geoscience Assessment project, 73
CUR. See Council for Undergraduate Research
CUSP. See Climate and Urban Systems Partnership
CYCLES program
  culturally congruent climate change education and, 22–23
  description of, 22
  evaluation of, 23–24
Cyclones, 3

Data
  arguments and, 35
  qualitative, 44–46
  quantitative, 46–47
DataStreme model, of in-service K-12 teacher professional development, 24–25
  course evaluation, 25
  courses, 24
  leadership training and renewal experiences for leaders, 25
  recruitment of teachers, 24–25
Deadlines, for science reporters, 113–114
Dean, Cornelia, 116
Debates, classroom, 36
Decision making
  learning about, K-12 science education framework and, 15
  learning outcomes and, 43
Democracy, closing information gap and, 119
Demographics, decision making and, 43
Descriptions, rich, K-12 science education framework and, 15
Design-based research, CUSP and, 108
Determinism, shift away from, 53
Dew points, increasing, 5
Digital technology, 6
Digital tools, CUSP and, 108
Disaster risk management, community-driven approaches to, 57
Dissemination, assessment and, 42
Diverse abilities of learners, assessment best practices and, 50
Diversity
  leveraging, participatory approaches and, 59
  REU programs and recruiting for, 79–80
*Don't Be Such a Scientist: Talking Substance in an Age of Style* (Olson), 116
Doughton, Sandi, 113
Drawings, 45–46, 45t
Droughts
  global climate change and, 5
  water availability in cities and, 103
Duration
  CYCLES and ICE-Net and, 22
  DataStreme courses and, 24
  teacher professional development and, 21
  Teachers as Watershed Researchers project and, 25

Earth. See also Civic understanding of planet Earth
  humanity as leading agent and engineer of, 1

Earth and Human Activity core idea, 19, 20
Earth and Space Science standards
  core ideas within, 19
  overarching intent of, 20
Earth hazards, 13
Earth science education
  K-12 science education framework and careers related to, 15
  core ideas, 11
  future of, 9
  interdisciplinarity of, 13
  teaching of, 16
  Literacy Principles developed for, 43
Earth Science Regents Exam (New York state), 43
Earth systems
  core ideas and, 19, 20
  expanded and elaborated view of, K-12 science education framework and, 14
  Grade 8 band end under, 11
Ecological knowledge
  indigenous systems of, 62
  traditional, pairing with western science, 63
Ecology, colonial history interwoven with, 62
Education. See also Earth science education; Geoscience education
  Anthropocene, multidisciplinary nature of, 6
  participatory approaches and, 61
Educational psychology, assessment instruments and, 51
Education levels, rise of, 6
Educators, obstacles in geoscience education for Native Americans and, 69
Einstein, Albert, 32
Emotion, learning integrated with, 90
Energy transfers and transformations, across Earth's systems, K-12 science education framework and, 12
Engineering, Practices and "inside" nature of, 10
Environmental education, scientific reasoning in, 33
Environmental journalism, reorganization at *New York Times* and, 111–112, 117
Environmental literacy, 41
Environmental science, assessment instruments and, 51
Epistemological reasoning, 36–37, 38
Equatorial regions, global warming and, 5

Essays, 45*t*
   essay-style questions, 44
   writing, 36
Ethics training, Anthropocene and, 83
European colonization, berries, traditional practices of Wabanaki people, and, 62
Evaluation
   CIRES Outreach and Education Group and, 28
   for civic understanding initiatives, importance of, 42
   of climate change professional development programs, 23–24
   CYCLES and ICE-Net and, 21, 22*t*
   of DataStreme courses, 25
   defining, 41
   Native-focused, for the geosciences, 73–74
   teacher professional development and, 21, 22*t*
   of Teachers as Watershed Researchers project, 26
Evaluative theory, 36
Evidence, 42
   coordinating theory and, 32
   K-12 science education framework and engaging in argument from, 12
Evidence-based argumentation, scientific reasoning and, 35–36, 38
Experiential learning, in informal science education institutions, 91
Experimental designs, generating, 32
Explanations, learning to make, K-12 science education framework and, 15
Explanatory journalism, hard news and, 112–113
Exploratorium, 100

Facebook, 6, 81, 114, 115
Facilitated dialogues, 57
Familial obligations, for Native American students in geoscience programs, 71
Ferris, Timothy, 111
Festival kits, CUSP and, 108
*Field Guide for Science Writers, A* (NASW), 111, 116
Field-tested Learning Assessment (FLAG) Guide, 48–49, 49*t*
Fieldwork, undergraduate geoscience education and, 77
Financial aid, for Native American students in geoscience programs, 71
Fisheries, maximum sustainable yield of, 3

"Five Ws and an H," journalists and, 114
Flathead Reservation, Pablo, Montana, 85
Flooding, in cities, 103
Florida Aquarium, 100
Focus groups, 45*t*, 46
Fond du Lac Band of Lake Superior Chippewa, 85
Fond du Lac reservation, Minnesota, summer 2012 flooding at, 72
Fond du Lac Tribal and Community College, 85
Forest clearing, atmospheric carbon dioxide and, 4
Formative evaluation, 41
Fossil fuels, burning of
   atmospheric changes related to, 4
   heat-trapping blanket analogy and, 90
*Framework for K-12 Science Education, A: Practices, Crosscutting Concepts, and Core Ideas* (NRC), 9. *See also* Earth science education
   DataStreme alignment with, 24
   reasoning and argumentation skills in climate education, 33
   vision and major dimensions of, 19
*Framework for 21st Century Skills*, 33
FrameWorks Institute, 90, 91, 95
Franklin Institute, 103
Funding, participatory approaches and, 61

GA. *See* Geoscience Alliance
Gelbspan, Ross, 118
GEMscholar Program, 86–87
Geoengineering, 4
Geologic record, pace of change and human activities unparalleled in, 5
Geoscience Alliance
   background of, 68
   future of, 74
   goals of, 68
   mission and vision of, 67
   national conferences, 68–69
   "Sustainability" Committee, 67
Geoscience Concept Inventory, 48, 49*t*
   multiple-choice questions from, 47
Geoscience education
   current assessment practices in, 47–49
   affect, 48–49
   cognition, 48
   history, advent of Next Generation Science Standards and, 28
   Native Americans and obstacles in at K-12 level, and role traditional knowledge can play in removing obstacles, 69–70

Geosciences
   culturally appropriate, Native-focused assessment and evaluation for, 73–74
   paradigm shifts in science and, 53
   removing barriers for Native Americans' participation in undergraduate and graduate education, 70–73
   completing geoscience degree, 71–72
   demographics of study sample, 70–71
   Salish Kootenai College hydrology degree program, 72–73
   underrepresentation of Native Americans in, 68, 74
Glacial melting, 3
Global climate change
   analogies and understanding of, 34
   increasing public's awareness about, 31
Global environmental consciousness, growth in, 6
Global warming
   cities and, 103
   hydrological cycle and, 5
   ocean temperature rise and, 3
   weather systems and, 5
GLOBE Program, 68
Graduate Record Examination (GRE), 78
Greenhouse effect, 36, 37
   "blanket analogy" as alternative explanation for, 34
   global warming and, 90
Greenhouse Effect Concept Inventory, 48
Greenhouse gases
   cities and emissions of, 103
   ocean acidification and increases in, 4
Grossman, Daniel, 116
Gross world production, 5–6
Groundwater formation, concept of scale and, 12
GWP. *See* Gross world production

H. J. Andrews Experimental Forest, Long-Term Ecological Research (LTER) Program, 25, 26
Harassment, Native American women in geoscience programs and, 71–72
Hard news, explanatory journalism and, 112–113
Hayes, Richard, 116
HBCUs. *See* Historically black colleges and universities

Healing, Native American community revitalization through, 70
Health issues, for Native American students in geoscience programs, 71
Heat energy, global warming and, 5
Heat engines, thermodynamics behind, 4–5
Heat events, extreme, cities and, 103
Heat-trapping blanket analogy, 90
Heat wave mortality
  city dwellers and, 104, 106
  social conditions associated with, 58
Henig, Robin Marantz, 116
Hierarchy of ways of knowing, 55
Hispanics, underrepresentation of, in the geosciences, 68
Historically black colleges and universities, 79
Holocene Epoch, 1
Houghton, Juliana, 113
Howes, Thomas, 72
Hubbert Curves, 12
Human activity, Earth substantially reconfigured by, 1
Human rights, climate and tribal displacement, just governance framework and, 63
Humans, Anthropocene and, 5–6, 53
Hurricane Katrina, 33, 53
Hurricane Sandy, 33
Hydrological cycle, global warming and, 5
Hydrologists, employment forecast for, 72
Hydrosphere, climatic system interconnections and, 33
Hypotheses, 32, 42

ICE-Net program
  culturally congruent climate change education and, 22–23
  description of, 22
  evaluation of, 23–24
Ice shelves, collapse of, 3
IMPACTS, ocean acidification research, 99–100
Income inequalities, 6
Indexing
  of common terms in text, 44
  simple and relational, 45
Indian Country, new generation of native scientists in, 69
Indian Ocean, fishing and energy extraction industries in, 3
Indigenous communities. *See also* Geoscience Alliance; Native Americans
  climate change and, 60, 61
    impacts, experiences and actions, 62–64

Indigenous knowledge, geoscience education and, 69
Indigenous worldviews, science-society gap and, 55
"Indigenuity," climate change adaptations and, 63
Industrial Revolution, atmospheric carbon dioxide and, 4
Informal reasoning, 32
Informal science education institutions, 92, 96
  communicating about ocean acidification at, 99–102
  with ocean theme or focus, 94
  public dialogue on climate change and, 90–91
Information gap, closing, democracy and, 119
Innovation, diffusion of, 91–92
Innovation, global digital technology and, 6
Inquiry-based instruction, advent of reforms focused on, 20
Institute for Broadening Participation, 79
Instrumentation design, assessment best practices and, 50
Interconnected world, challenges of the Anthropocene and, 6
Interdepartmental relationships, Native American students in geoscience programs and, 71
Interdisciplinarity, of Anthropocene and Earth Sciences, K-12 science education and, 13
Interdisciplinary teams, K-12 science education framework and, 17
Intermountain Climate Education Network. *See* ICE-Net
Internal combustion engine, 4, 5
International Association for Geoscience Diversity, 80
*International Business Times*, 112
International Panel on Climate Change, 114, 117
Internet, 6, 115
Interpretive leaders, public awareness of climate change and, 92
Interrater reliability, 45
Interviews, 45t, 46
Invasive species, wild rice, White Earth Nation and, 60
IPCC. *See* International Panel on Climate Change
IQs, global rise in, 5
ISEIs. *See* Informal science education institutions

Jeffersonian Science, 56
Jordan, Chris, 61
Journalism
  coverage of climate change story and, 117–118
  explanatory, hard news and, 112–113
Journalists
  climate change coverage and skills for, 118
  scientists and, 111
Justice, climate and tribal displacement, governance framework and, 63

Kelly, Walt, 118
Kepler, Johannes, 34
Knowledge-building activities epistemological reasoning and, 37
Knowledge in community, valuing, 58
Knudson, Mary, 116
K-12 science education framework
  Anthropocene and, 9–17
  conclusion, 16–17
  critical framework elements, 10–11
  genesis and grounding of, 9–10
  overall goal, 10
  ten recommendations for designing standards and curriculum, 11–16

Land, Anthropocene and impact on, 2
Land loss, in the Louisiana bayou, 60
Landscape formation, concept of scale and, 12
Large-scale revisions, K-12 science education framework and, 17
Learning
  assessment data and, 42
  meaningful, analogies and, 34
  outcomes
    affect and, 43
    assessment approaches and, 44
    Backward Design and, 49–50
    categorizing, 42–43
    cognition and, 43
    decision making and, 43
    identifying, 42
  participation and, 106, 107
  styles, obstacles in geoscience education for Native Americans and, 69–70
Leech Lake Tribal College, 87
Leopold Leadership Program, 115
Life expectancy, global, increase in, 5
Life Sciences, K-12 science education framework and teaching of, 16
Likert, fixed response assessment, summary of, 47t
Likert, sliding scale response, summary of, 47t

Likert-type item, from New Ecological Paradigm Scale, 46
Likert-type questions, 46–47
Literacy Principles, for Earth, Ocean Atmosphere, and Climate sciences, 43
Lithosphere, climatic system interconnections and, 33
"Loading dock" model of science, 55
Local contexts, K-12 science education framework and transferability of applications in, 13
Louisiana bayou, land loss in, 60
Louis Stokes Alliance for Minority Participation program, 79
Low-carbon living, cities and, 103
Lubchenco, Jane, 99, 102

Marian Koshland Science Museum, 103
Marine environments, coastal human populations and, 3
Mathematics skills, Native American students in geoscience programs and, 71
Matter transfers and transformations, across Earth's systems, K-12 science education framework and, 12
McCarthy, Gina, 117
McKibben, Bill, 118
McNair programs, 79, 80
Meaning-making, understanding climate change and, 90
Media
 foundation-shaking changes in, 111, 112
 resources for scientists dealing with, 115–116
 scientists working in, tips for, 113–115
Mediarology website, 116
Meningitis, in the Sahel, research project, 59–60
Meningits and environmental risk information technology, 60
Mental health issues, for Native American students in geoscience programs, 71
Mentoring, co-learning and, 69
Mentors, REU programs and communication between and among, 81–82
MERIT. See Meningits and environmental risk information technology
MESA program, weather and water after-school student-scientist project: waterspotters, 27
Meteorological conditions, meningitis in the Sahel and, 59

Meyer, Philip, 116
Minnesota, global carbon dioxide releases and climate in, 5
Minorities
 dearth of, in science, 53, 54
 recruiting for undergraduate research opportunity programs, 79–80
 underrepresentation of, in the geosciences, 68
Minority-serving institutions, 78, 79, 80
Model-based epistemology, 36
Mode 2 science, 56
Morality, decision making and, 43
MOTE Marine Laboratory and Aquarium, 100
MSIs. See Minority-serving institutions
Multiple-choice, multiple response test, 47t
Multiple-choice, single response test, 47t
Multiple-choice questions
 assessment and, 47
 from Geoscience Concept Inventory, 47
Multiple priorities, embracing, 58
Multiplist theory, 36

NAAEE. See North American Association for Environmental Education
NAEP. See National Assessment of Educational Progress
NASA, 67, 68, 74
NASA Innovations in Climate Education, 22
NASW. See National Association of Science Writers
National Academy of Sciences, 53, 103
National Aquarium, 100
National Assessment of Educational Progress, 43, 48, 49t
National Association for Research in Science Teaching, 25
National Association of Science Writers, 111, 116
National Center for Atmospheric Research, 60, 84, 85
National Center for Earth-surface Dynamics, University of Minnesota, 68, 85, 86
National Council for Geographic Education, 25
National Earth Science Teachers Association, 25
National Fish and Wildlife Service, Columbia River Inter-Tribal Fish Commission and, 63
National Institutes for Health, 77

National Lacustrine Core Facility, University of Minnesota, 85
National Marine Educators Association, 25
National Network for Ocean and Climate Change Interpretation, 91, 92–95
 activities and outcomes anticipated by, 92
 informal science education institutions with ocean theme, 94
 legacy and sustainability, 95–96
 logic model, 93
 results to date, 95
 social capital and, 91–92
 target audience, 92, 95
National Oceanic and Atmospheric Administration, 63, 74, 77, 84, 99
National Research Council, 77
*National Science Education Standards* (NRC), 10, 20
National Science Foundation, 68, 74, 96
 Cultural Validity in Geoscience Assessment project funded by, 73
 GEMscholar Program and, 87
 Geoscience Directorate, 77
 REU on Sustainable Land and Water Resources and, 86
 SOARS Program funding through, 85
 WEBCASPAR site, 79
National Science Teachers Association, 25
 Anchors Project, 10
National strategy on climate change, 91–92
 climate scientists and, 91
 communities of practice and, 91
 informal science education practitioners and, 91
 interpreters as communication strategists and, 91
 learning researchers and, 91
 social networks and diffusion of innovation and, 91–92
Native Americans. See also Undergraduate research opportunity program
 culturally appropriate, Native-focused assessment and evaluation for, 73–74
 GEMscholar Program for, 86–87
 in geosciences, broadening participation of, 67, 68
 as learners, common attributes of, 70
 obstacles in geoscience education for, and role traditional knowledge can play in removing obstacles for, 69 70

128   INDEX

Native Americans. *See also* Undergraduate research opportunity program (*cont'd*)
  removing barriers for participation in undergraduate and graduate education in the geosciences, 70–73
    completing geoscience degree, 71–72
    demographics of study sample, 70–71
    Salish Kootenai College hydrology degree program, 72–73
  Research Experiences for Undergraduates, inclusiveness, and, 78
  underrepresentation of, in the geosciences, 68, 74
Native Hawai'ians, traditional knowledge in geoscience education and, 69
Natural disasters, science-society gap and vulnerability to, 53–54
Natural gas, carbon in, 4
Natural phenomena, epistemological representations and, 36
Natural systems, K-12 science education framework and teaching about, 13
Nature Conservancy, The, 119
Navajo community, Cultural Validity in Geoscience Assessment project and, 73
NCAR. *See* National Center for Atmospheric Research
NCED. *See* National Center for Earth-surface Dynamics
Neighborhood programming, CUSP and, 108
*New Ecological Paradigm* (NEP) Scale, 23
New England Aquarium (NEq), 91
New Knowledge Organization Ltd., 91
*News & Numbers: A Guide to Reporting Statistical Claims and Controversies in Health and Related Fields* (Cohn), 116
Newspapers, declining science sections in, 112
Newsroom employment, decline in, 112
News sources, science reporters and, 113–115
New York City, engagement around coastal flood risk in, examples of, 107
New York Hall of Science, 103
New York State Regents Exams, 48, 49*t*
*New York Times*, climate science and newsroom reorganization at, 111–112, 117
*New York Times Reader, The: Science & Technology* (Stocking, et al.), 116

Next Generation Science Standards (NRC), 10, 21, 24, 28
NICE. *See* NASA Innovations in Climate Education
NIH. *See* National Institutes for Health
Nitrogen, atmospheric, 4
NNOCCI. *See* National Network for Ocean and Climate Change Interpretation
NOAA. *See* National Oceanic and Atmospheric Administration
Nonpersuasive communication strategy, understanding climate change and, 90
Nontraditional students, supporting in REU programs, 80–81
North America, biomes of, 2
North American Association for Environmental Education, 49, 49*t*
North Carolina Aquarium, Fort Fisher, 100
North Pacific Landscape Conservation Cooperative, tribal input to climate-related changes in forests and, 63
Northwest Commission of Colleges and Universities, 72
NRC. *See* National Research Council
NSTA. *See* National Science Teachers Association

Observations, 45*t*, 46
Ocean acidification, 89
  carbon dioxide emissions and, 3–4
  communicating about ocean acidification at ISEIs, 99–102
    implications for, 101–102
    key findings, 100–101, *101*
    methodology, 99–100
Ocean literacy, 89
Oceanography Society, The, 92
Ocean Project
  methodology used by, 99–100
  public opinion research by, 99
Oceans
  Anthropocene and, 2–4
  humanity and pH alterations in, 4
Ocean science, Literacy Principles developed for, 43
Ohio State University, 91
Oil, carbon in, 4
Oil-drilling platforms, 3
Olson, Randy, 116
Online communities, public awareness of climate change and, 96
Open-ended questions, 44
Oppenheimer, Michael, 113
Oregon Coast Aquarium, 100

Oregon Natural Resources Education Program, Oregon State University, 25, 26
Overfishing, 3
Oxygen, atmospheric, 4

Pacific Northwest, tribal subsistence and migratory salmon in, 62–63
Pacific Ocean, fishing and energy extraction industries in, 3
Pacific Science Center, 100
Participation, learning and, 106, 107
Participatory approaches
  education and, 61
  funding and, 61
  review process and, 61
  tenure and promotion and, 61
Participatory projects
  land loss in the Louisiana bayou, 60
  meningitis in the Sahel, 59–60
  successful, 58–59
  wild rice and White Earth Nation, 60–61
Participatory research, community-driven, 58
Pastures, 2
Pattern analysis, computer-based, of drawings, 46
Patterns, as crosscutting concept in science, 12
PEAR. *See* Program in Education, Afterschool, and Resliency
PEARweb, 48, 49*t*
Pedagogical approaches, novel, traditional knowledge incorporated with, 70
Peer learning, SOARS program and, 84
Pennsylvania State University, 91
Personal essays, REU program applications and, 80
pH, shells of marine creatures, corals and, 4
Phenomenon-based epistemology, 37
*Phoenix, The* (Boston), demise of, 117
Photosynthesis, atmospheric carbon dioxide and, 4
Physical communities, CUSP and, 107, 108
Place, sociocultural landscape for Native people and, 73
Place-based undergraduate research programs, creating, 83–84
Planetary sustainability insurance, 34
Plants
  human activities, species loss and, 2
  photosynthesis, atmospheric carbon dioxide and, 4
Podcasts, 119

Polar regions, warming of, 5
Pollution, coastal human populations and, 3
Pooley, Eric, 118
Population, percentage of, along coastlines, 2–3
Portland, Oregon, citizens' involvement in watershed management in, 105
Post-normal science, 56
PPSR. *See* Public Participation in Scientific Research
Practices, K-12 science education framework
 case studies in, 15
 designing standards and curriculum, recommendations for, 12
 presentation of, 10
 progressions of teaching and learning, 16
Pre-assessments, 41
Precipitation, increasing, water vapor in atmosphere and, 3
*Precision Journalism: A Reporter's Introduction to Social Science Methods* (Meyer), 116
Predictions, K-12 science education framework and, 15
Professional development evaluation, critical levels of, 22t
Program in Education, Afterschool, and Resiliency, assessment instruments, 48
Project Atmosphere workshop, 25
Psychometrics, assessment instruments and, 51
Public dialogue on climate change, 89–96
 developing a national strategy, 91–93
 general public awareness of climate change, 89
 cognitive and social psychology of climate change, 89–90
 National Network for Ocean and Climate Change Interpretation, 92–95
 informal science education institutions with ocean theme, 94
 legacy and sustainability, 95–96
 logic model, 93
 results to date, 95
 target audience, 92, 95
 potential of informal science education, 90–91
Public health, community-driven approaches in, 57
Public Participation in Scientific Research, 57
Purdue University, 68

Pyramid Lake Paiute tribe, climate change, water shortages and, 62

Qualifiers, arguments and, 35
Qualitative assessments
 drawings, 45–46, 45t
 interviews and focus groups, 45t, 46
 observations, 45t, 46
 short response questions and essays, 44–45, 45t
 summary of, applicable to educational settings, 45t
Quantitative assessments, 46–47
 Likert-type questions, 46–47
 multiple-choice questions, 47

Rainfall events, extreme, 5, 103
Reasoning
 with analogies, 34–35, 37–38
 definition and description of, 31–32
 epistemological, 36–37
 scientific
  climate literacy enhanced with, 33–37, 38
  defined and described, 31–32
  importance in climate education, 32–33
Rebuttals, arguments and, 35, 36
Recruiting
 American Indian students for GEMscholar Program, 86–87
 for undergraduate research opportunity programs, 79
Red Lake Nation College, 87
Reductionism, shift to systems thinking from, 53
Relation-based epistemology, 36, 37
Reporters. *See* Journalists; Science reporters
Research
 assessment, 41
 Earth Science education and importance of, 16
 excellence, science-society gap and, 55
Research experiences
 teaching students about Anthropocene in, 83–84
 ethics training, 83
 place-based REU, working with communities, 83–84
Research Experiences for Undergraduates
 on Sustainable Land and Water Resources, 85–86
 value of, for promoting learning in geosciences, 77

Research question, collaborative definition of, 57–58
Research teams, REU programs and, 82
Resource management practices, traditional knowledge systems and, 69
REUs. *See* Research Experiences for Undergraduates
Review process, participatory approaches and, 61
Revkin, Andrew C., 116, 188

SACNAS. *See* Society for the Advancement of Chicanos and Native Americans in Science
Sagan, Carl, 53
Sahel, meningitis in, research project, 59–60
Salish Kootenai College (Montana), 68, 85, 86
 hydrology degree program at, 72–73
 two- and four-year geoscience degree programs at, 72
Salmon-dependent resources, tribal subsistence in Pacific Northwest and, 62–63
Sanjayan, M., 119
Savannas, 2
Scaffolding, undergraduate research projects, 82
Scale, as crosscutting concept in science, 12
Schneider, Stephen H., 115, 116
Science. *See also* Community-driven science
 dearth of minorities in, 53, 54
 evidence and, 42
 extended use of history, philosophy, and sociology of, K-12 science education and, 14–15
 "loading dock" model of, 55
 paradigm shifts in the Anthropocene, 53
 pattern as crosscutting concept in, 12
 Practices and "inside" nature of, 10
 scale as crosscutting concept in, 12
Science education, scientific reasoning in, 32, 33
*Science for All Americans* (Rutherford and Ahlgren), 10
Science literacy, 19, 43
Science Museum of Minnesota, 100
"Science push" model, of science-policy interaction, 55
Science reporters
 climate change coverage and skills needed by, 118
 tips for, 113–115

130  INDEX

Science-society gap
 bridging, 54–55
 closing, 55–57, 56
 geosciences and evidence of, 53
*Science the Endless Frontier*, 55
Science writing, great, requirements for, 111
Scientific claims, evidence-based argumentation and, 35
Scientific inquiry, on Tibetan Plateau, 27
Scientific method, responsible media and, 116
Scientific reasoning
 climate education and importance of, 32–33
 climate literacy enhanced with, 33–37
  epistemological reasoning, 36–37
  evidence-based argumentation, 35–36
  reasoning with analogies, 34–35
 as crucial cognitive mechanism, 37
 definition and description of, 31–32
 epistemological representations and, 36
 everyday thinking and, 32
Scientists
 journalists and, 111
 resources for media dealing with, 115–116
 working in media, tips for, 113–115
*Scientists' Guide to Talking with the Media, A: Practical Advice from the Union of Concerned Scientists*, 116
Sea level rise
 cities and, 103
 climate change and, 89
 land loss in the Louisiana bayou and, 60
 warming ocean waters and, 3
Seattle Aquarium, 100
*Seattle Times*, 113
*Second Assessment Report of the Intergovernmental Panel on Climate Change* (IPCC), 32
*SEJournal*, 119
Self-efficacy belief, climate change education and, 105
Semi-quantitative data, 46
Shabecoff, Phil, 118
Shifting Baselines Ocean Media Project, 116
Short response questions and essays, 44–45, 45t
Significant Opportunities in Atmospheric Research and Science Program, 84–85
"Six Americas" survey, 23
SKC. *See* Salish Kootenai College

Skepticism, science-society gap and, 54
Skills, cognition and, 43
SOARS Program. *See* Significant Opportunities in Atmospheric Research and Science Program
Social activism, 95
Social capital, NNOCCI and, 91–92
Social development, learning integrated with, 90
Social media, 114, 115, 119
Social networks, 91–92
Social psychology, of climate change, 89–90
Social sciences, K-12 science education framework and
 extended use of, 13–14
 teaching, 16
Social systems, K-12 science education framework and teaching about, 13
Society. *See* Science-society gap
Society for the Advancement of Chicanos and Native Americans in Science, 68, 79
Society of Environmental Journalists, 119
Socioscientific reasoning, dimensions within, 32
Solar energy, atmosphere and differences in input with, 5
Solutions-oriented science, 56
Southern Louisiana Wetlands Discovery Center, 60, 85
Soybeans, 2
St. Anthony Falls Laboratory (SAFL), 85
Stanford Woods Institute for the Environment, 115
Steinhart Aquarium (California Academy of Sciences), 100
STEM
 curriculum, traditional knowledge coupled with current science curriculums and, 70
 disciplines, traditional knowledge incorporated into geoscience education and, 69
 learning outcomes, evaluation of DataStreme courses and, 25
 professionals, highly qualified, developing, 19
 programs, undergraduate geoscience research and, 77, 78
Stocking, S. Holly, 116
Storms, warming oceans and formation of, 3
Student learning, assessment of, evidence and, 42

Students with disabilities, attracting, for REU programs, 80
Study circles, public awareness of climate change and, 95
Summative evaluation, 41
Sun-Earth connections education and outreach, extreme ultraviolet variability experiment geomagnetism in MESA classroom, 27
Survey questions, written responses to, 44
Sustainability, K-12 science education framework and, 16
Systems thinking
 CUSP programming and, 106–107
 shift from reductionism to, 53

Teacher professional development in the Anthropocene, 19–28
 active learning, 21
 coherence, 21
 connecting geoscientists and educators, 26–28
  climate and oceans teacher professional development, 27
  scientific inquiry on the Tibetan Plateau, 27
  strategies for success, 28
  Sun-Earth connections education and outreach, 27
  weather and water after-school student-scientist project, 27
 degree of collective participation, 21
 duration, 21
 examples of programs for, 21–26
  American Meteorological Society DataStreme Model, 24–25
  climate change education in American Indian communities, 22–23
  evaluation of climate change professional development programs, 23–24
  teachers as watershed researchers project, 25–26
 focus on content, 20–21
 program evaluation, 21
Teachers, obstacles in geoscience education for Native Americans and, 69
Teachers as Wateshed Researchers (TWR) project, 25–26
Technology, global connectivity via, 6
Technology gap, in Native American reservation communities, 74
Temperature, atmosphere and differences in, 5
Temporal communities, CUSP and, 107, 108

Tenure and promotion, participatory approaches and, 61
Thematic content analysis, 45
Theory, coordinating evidence and, 32
Tibetan Plateau, scientific inquiry on, 27
Toulmin's argumentation pattern, 35
Town-hall meetings, 57
Trace gases, atmospheric, 4
Traditional knowledge
 environmental/ecological, 69
 incorporating into geoscience education, 67, 69–70
Transportation, coastal flooding and, 103
Tribal colleges and universities, two- and four-year geoscience degree programs at, 72
Tribal communities, climate change and, 62–64
Tribal lands, Geoscience Alliance and resource-management decisions on, 67
Trio program, 80
Tropical rainforests, 2
Trust building, participatory approaches and, 69
Trusted messengers, informal science education institutions and, 102
Twitter, 114, 115

UCAR. See University Consortium for Atmospheric Research; University Corporation for Atmospheric Research
ULNs. See Urban Learning Networks
Uncertainty
 in science, role of, 42
 science-society gap and, 54
Undergraduate research opportunity program
 inclusive, considerations for planning, 78–82
  advertising and recruiting, 79
  application for, 80
  attracting students with disabilities, 80
  communication, socialization and, 81–82
  goal of program, 78
  identify the ideal participant, 78–79
  multiple advisors and team-structured collaborative research, 82
  nontraditional student support, 80–81
  orientation and final week, 81
  recruiting for diversity, 79–80
  scaffolding research experiences, 82
  selection for, 80
Undergraduate Research Program, National Science Foundation and, 77
Underrepresented groups, SOARS program and, 85
Underserved community members, Research Experiences for Undergraduates, inclusiveness, and, 78
United States, coastal population of, 89
Universalism, science-society gap and, 54
University Consortium for Atmospheric Research, 68
University Corporation for Atmospheric Research, 84, 85
University of Colorado, Boulder, 84, 85
University of New Orleans, 85
University of Pittsburgh Learning Research and Development Center, 103
University of Washington, 113
Urban climate education
 city-wide collaborations for, 103–108
  CUSP, 103, 105–108
  Urban Learning Networks, 105–106, 107, 108
Urban heat island effect, 103
Urban Learning Networks, 105–106
Urban population
 climate change and, 102–103
 global, 103
Urban systems, CUSP programming and, 107
URP. See Undergraduate Research Program
U.S. Commission on Civil Rights, 69
Use-inspired science, 56
US Geological Survey (USGS), 74, 77

Vanishing Points app, land loss in the Louisiana bayou and, 60
Vector-borne disease, cities, climate change and, 103
Virginia Aquarium and Marine Science Center, 100
Virtual communities, CUSP and, 107
Volunteerism, societal impacts of, 95

Wabanaki people (Maine and Canada), European colonization, berries and traditional practices of, 62
Warrants, arguments and, 35
Waterborne microbial health risks, Crow Tribe, NOAA data and mitigation of, 63

Water pollution, flooding in cities and, 103
Water-resource hazards, tribal communities in U.S. and, 63
Water resources, climate change and stresses on, 72
Watershed researchers, teacher professional development project, 25–26
Water vapor in atmosphere, increased precipitation and, 3
Watts, Bueno, 70
Weather and Water after-school student-scientist project: waterspotters, 27
Weather systems, global warming and slowing circulation of, 5
WEBCASPAR site (NSF), 79
Websites, 115
Western science
 perspectives, Native Americans in geoscience programs and, 72
 traditional ecological knowledge paired with, 63
White Earth Nation, wild rice and, 60–61
White Earth Natural Resources Department, 60
White Earth Tribal and Community College, 60
Whitehat, Albert, 55
Wildcat, Daniel, 63
Wild rice (or manoomin)
 Anishinaabeg people of Great Lakes region and, 60
 hydrologic events and impact on, 72
 White Earth Nation and, 60–61
Women
 Native American, geoscience programs and harassment of, 71–72
 successful participatory approaches and, 69
Woods Hole Oceanographic Institute, 91, 96
Writing, well about science, requirements for, 111

Young scientists, public awareness of climate change and, 92
Youth interpreters, public awareness of climate change and, 92

Zara, Christopher, 112
Zimmer, Carl, 116
Zoos, Ocean Project in partnership with, 99